A FIRST COURSE ON ELECTRICAL DRIVES

(Second Edition)

S.K. Pillai
Professor
Department of Electrical Engineering
Indian Institute of Technology
Bombay

JOHN WILEY & SONS
New York Chichester Brisbane Toronto Singapore

First Published in 1989 by
WILEY EASTERN LIMITED
4835/24 Ansari Road, Daryaganj
New Delhi 110 002, India

Distributors:

Australia and New Zealand:
JACARANDA-WILEY LTD., JACARANDA PRESS,
JOHN WILEY & SONS, INC.
GPO Box 859, Brisbane, Queensland 4001, Australia

Canada:
JOHN WILEY & SONS CANADA LIMITED
22 Worcester Road, Rexdale, Ontario, Canada

Europe and Africa:
JOHN WILEY & SONS LIMITED
Baffins Lane, Chichester, West Sussex, England

South East Asia:
JOHN WILEY & SONS, INC.
05–05 Block B, Union Industrial Building
37 Jalan Pemimpin, Singapore 2057

Africa and South Asia:
WILEY EASTERN LIMITED
4835/24 Ansari Road, Daryaganj
New Delhi 110 002, India

North and South America and rest of the world:
JOHN WILEY & SONS, INC.
605 Third Avenue, New York, N.Y. 10158, USA

Library of Congress Cataloging in Publication Data

ISBN 0-470-21399-X John Wiley & Sons, Inc.
ISBN 81-224-0166-X Wiley Eastern Limited

Printed in India at Prabhat Press, Meerut. U.P.

To

Velandi Sivan Pillai

AND

Krishnammal Sivan Pillai

my parents,
who gave everything they had to
make me as I am now

Foreword

Electrical Drives play a vital role in engineering and industry both in this country and abroad. It is therefore essential that students of electrical engineering have a proper grounding in this subject. Conventional courses in Electrical Machines, however, are not adequate for the purpose as electric motors do not by themselves constitute an electrical drive and their characteristics have to be studied keeping in mind the types of control schemes (such as those using thyristor circuits) and the dynamics of the load. On the other hand, courses on 'Control Systems', 'Industrial Electronics' and 'Power Electronics' do not devote sufficient attention to electrical motor characteristics and mechanical load demands. It is thus necessary to have a course on the fundamentals of electrical drives, suitable for study by undergraduate students of electrical engineering. This book *A First Course on Electrical Drives*—is designed to meet the need for a textbook in English for such a course. The writing of the book has been supported by the Curriculum Development Cell of the Indian Institute of Technology, Bombay.

The book gives a comprehensive introduction to the dynamics of drives, the characteristics, starting and braking of dc and ac motors, as also their loading conditions, ratings and heating. There are separate chapters devoted to solid state controlled drives and industrial applications. The MKS system of units has been used throughout and Indian Standards Specifications have been adhered to. In addition to worked examples, most chapters include a number of problems designed to test the student's grasp of the subject.

The author, Dr. S.K. Pillai, has over twenty years experience of teaching and research in electrical engineering, and he has developed the material of this book over the past ten years while conducting lecture, tutorial and laboratory classes for final year undergraduate students of Electrical Engineering at the Indian Institute of Technology, Bombay. The style and organization of the work reflects the discerning insight of a teacher into the requirements of a student and each topic is developed step by step in a clear and cogent manner. I am confident, therefore, that this book will be welcomed by students and teachers alike.

April 16, 1982

R.E. BEDFORD
Dy. Director
Indian Institute of Technology
Bombay

Foreword

[illegible] has a vast role in engineering and industry both [illegible] technology, [illegible]

The book [illegible]

[illegible]

Director

Preface to the First Edition

Electrical drives offer a convenient means for controlling the operation of different equipment used in industry. The high reliability and great versatility of electrical drives, especially of those controlled by solid state devices, have resulted in their wide application. In fact, the growth and developments of electrical drives have been closely in parallel with those of automation in industry. In this context, it is indeed surprising that the theory and practice of electrical drives is even, at present, taught for Bachelor's degree students as one amongst various topics under the head "Utilization of Electrical Energy" and not as a separate subject, in many technical institutions in our country. One of the reasons may be the non-availability of suitable text books on the subject in English language. This volume is primarily written with the aim of providing a text book for the undergraduate students in Electrical Engineering on the fundamentals of electrical drives.

The contents of this book have gradually evolved over the last ten years from the notes used by the author in teaching final year undergraduate students of I.I.T. Bombay. A number of books and papers published, listed in the 'Bibliography' have been made use of in the preparation of this book. They not only indicate the extent of indebtedness of the author to those authors and publishers, but also provide the student with material for further reading. Most chapters include, in addition to a number of solved examples, problems to test how well the student has grasped the subject matter presented.

I gratefully acknowledge the assistance and encouragement given to me by several of my past students and colleagues. A special word of thanks is due to Prof. R.E. Bedford, who kindly agreed to write a Foreword for the book. I am very much indebted to the Curriculum Development Cell, I.I.T. Bombay for giving me the necessary financial support to prepare the first version of this book. Grateful acknowledgement is due to the Indian Standards Institution for granting permission to use the definition and the associated figures relevant to the different classes of duty indicated in I.S.S. 4722. I also wish to thank the production department of Wiley Eastern Limited for the great care with which they have worked in bringing the book to its present form. Last but not least, I wish to record my gratitude to my wife Vijaya and sons Sivan and Kumar for the sacrifices they have made that this work could be completed.

I.I.T. Bombay
April 1982

S.K. PILLAI

Preface to the Second Edition

The present edition, essentially, introduces the students of Electrical Engineering to the recent additions in solid state controlled drives viz., brushless d.c. motors and switched reluctance motors.

In response to the comments made on the first edition by many teachers, a separate chapter on 'Stepper Motors' has been included. The drives used in the fast growing Indian Petroleum industry have also been discussed, in the chapter on 'Industrial Applications'.

The author would like to acknowledge the valuable advice and suggestions of many instructors and students who used the earlier edition.

I.I.T. Bombay
August 1988.

S.K. PILLAI

Contents

Chapter 1

Introduction

1.1 Concept of an Electrical Drive

Most of the production equipment used in modern industrial undertakings consist of three important components, namely, the prime mover, the energy transmitting device and the actual apparatus or equipment that performs the desired job. The function of the first two components is to impart motion and operate the third one. The most commonly used prime mover is, of course, an electric motor, since it is far superior in performance to steam, hydraulic, diesel and other types of engines. Electric motors are, often, operated directly from a supply line, under their own inherent speed-torque characteristics and their operating conditions are dictated by the mechanical loads, connected to them. However, in many applications, the motors are provided with a control equipment by which their characteristics can be adjusted and their operating conditions with respect to the mechanical load varied to suit specific requirements. The most common control adjustment is of motor speed, but torque and acceleration or deceleration can also be adjusted. The control equipment usually consists of relays, contactors, master switches and solid state devices such as diodes, transistors and thyristors.

The aggregate of electric motor, the energy transmitting shaft and the control equipment by which the motor characteristics are adjusted and their operating conditions with respect to mechanical load varied to suit particular requirements, is called an *electrical drive*. The drive together with the load constitutes the drive system.

1.2 Classification of Electrical Drives

In general, electrical drives may be classified into three categories: Group drive, Individual drive and Multimotor drive.

Group drive consists of a single motor which actuates several mechanisms or machines by means of one or more line shafts supported on bearings. It is also called a line shaft drive. The line shaft fitted with multistepped pulleys and belts that connect these pulleys and the shafts of the driven machines serve to vary their speed.

Even after taking into account the cost of line shafts, pulleys, belts and other installations, the group drive is the most economic one, since the rating of the motor used may be comparatively less than the aggregate of ratings of the individual motors required to drive each equipment, because all of them may not be working simultaneously.

But, seldom is the group drive used, nowadays, due to the following disadvantages:

(a) Any fault that occurs in the driving motor renders all the driven equipment idle.
(b) Considerable power loss takes place in the energy transmitting mechanisms.
(c) Flexibility of layout of the different machines is lost, since they have to be located as to suit the layout of the line shaft.
(d) The use of line shaft, pulleys and belts make the drive untidy in appearance and less safe to operate.
(e) The level of noise produced at the worksite is quite high.

In the individual drive, an electric motor is used for transmitting motion to various parts or mechanisms belonging to a single equipment. For example, such a drive in a lathe rotates the spindle, moves the feed and also with the help of gears, imparts motion to the lubricating and cooling pumps of the lathe. In many applications, the individual drive consists of a motor, which is specially designed to form an integral part of the equipment.

In the case of individual drive too, the energy is transmitted to the different parts of the same mechanism by means of mechanical parts like gears, pulleys etc. Hence, there occurs some power loss. This disadvantage is removed in the case of multimotor drives.

In multimotor drives, separate motors are provided for actuating different parts of the driven mechanism. For example, in travelling cranes, there are three motors: one for hoisting, another for long travel motion and the third for cross travel motion. Paper mills, rolling mills, rotary printing machines, metal working machines etc. employ a large number of multimotor drives.

The use of individual drives and multimotor drives has enabled introduction of automation in production processes, which in turn has considerably increased the productivity of different industrial undertakings. Complete or partial automation helps to operate various mechanisms at optimum conditions and to increase reliability and safety of operations.

Chapter 2

Dynamics of Electrical Drives

The electromagnetic forces or torques developed in the driving motor tend to propagate motion of the drive system. This motion may be uniform if the linear velocity (in the case of translational motion) or the angular velocity (in the case of rotational motion) is constant, or non-uniform, as it occurs while starting, braking or changing the load on the drive.

In case of uniform motion the torque developed by the driving motor is to overcome any resisting torque offered by the driven equipment as well as the torque due to friction. In other words, only static resisting torques, commonly called as load torques, are to be counterbalanced, if the motion were uniform.

2.1 Types of Loads

Loads can be of two types—those which provide active torques and those which provide passive torques.

Active torques are due to either gravitational force or deformation in elastic bodies. The active torques due to gravitational pull are obtained in case of hoists, lifts or elevators and railway locomotives operating on gradients. Such torques are also developed during compression or release of springs. Since the functioning of hoisting mechanisms, operation of locomotives on gradients and compression or release of springs are all associated with a change in potential energy of the drive, active torques are also closely connected to the potential energy. When a load is moved upwards or a spring is compressed, the stored potential energy increases and the active torque developed opposes the action that takes place, i.e., the torque is directed against the upward movement or compression. On the other hand, when a load is brought downwards or a spring is released the stored potential energy decreases and torque associated with it aids the action. Thus, it can be seen that the active torques continue to act in the same direction even after the direction of the drive has been reversed.

Passive torques are those due to friction or due to shear and deformation in inelastic bodies (lathes, fans, pumps etc.). They always oppose motion, retarding the rotation of the driven machine. Moreover, with change in direction of motion, the sense of torque also changes. For example, when a weight is being lifted up, the friction torque adds to the useful torque, but when lowered down it subtracts from the latter.

2.2 Quadrantal Diagram of Speed-Torque Characteristics

In view of the fact that both active and passive load torques can be present in general, in a drive system, the motor driving the load may operate in different regimes—not only as a motor, but for specific periods, also as a generator and as a brake. Further, in many applications, the motor may be required to run in both directions. Therefore, in sketching the speed torque characteristics of either the load or the motor, it is preferable to use all four quadrants of the speed-torque plane for plotting, rather than to confine the characteristics to the first quadrant alone. When drawn in this manner, the diagram is referred to as quadrantal digaram.

The conventions used for positive and negative values of speed, motor torque and load torque in a diagram of this type must be understood very clearly. The speed is assumed to have a positive sign, if the direction of rotation is anticlockwise or is in a such a way to cause an 'upward' or 'forward' motion of the drive. In the case of reversible drives, the positive sign for speed may have to be assigned arbitrarily either to anticlockwise or clockwise direction of rotation.

The motor torque is said to be positive if it produces an increase in speed in the positive sense. The load torque is assigned a positive sign when it is directed against the motor torque.

Figure 2.1 shows the four quadrant operation of a motor driving a hoist consisting of a cage with or without load, a rope wound onto a drum to hoist the cage and a balance weight of magnitude greater than that of the empty cage but less than that of the loaded one. The arrows in this figure indicate the actual directions of motor torque, load torque and motion in the four quadrants. It can be easily seen that they correspond to the sign conventions stated earlier for speed, motor torque and load torque.

The load torque of the hoisting mechanism may be assumed to be constant, i.e., independent of speed, since the forces due to friction and windage are small in the case of low speed hoists and the torque is primarily due to the gravitational pull on the cage. This torque being an active load torque doesn't change its sign even when the direction of rotation of the driving motor is reversed. Therefore, the speed torque curves of a hoist load can be represented by means of vertical lines passing through two quadrants. The speed torque characteristic of a loaded hoist is shown in Fig. 2.1 by means of the vertical line passing the first and fourth quadrants. Since the counterweight is assumed to be heavier than the empty cage, the inherent tendency of the load, viz., the empty cage is to move in an opposite direction to that of load presented by the loaded cage and hence the speed torque curve of the unloaded hoist is represented by the vertical line passing through second and third quadrants.

In the first quadrant the load torque acts in a direction opposite to that

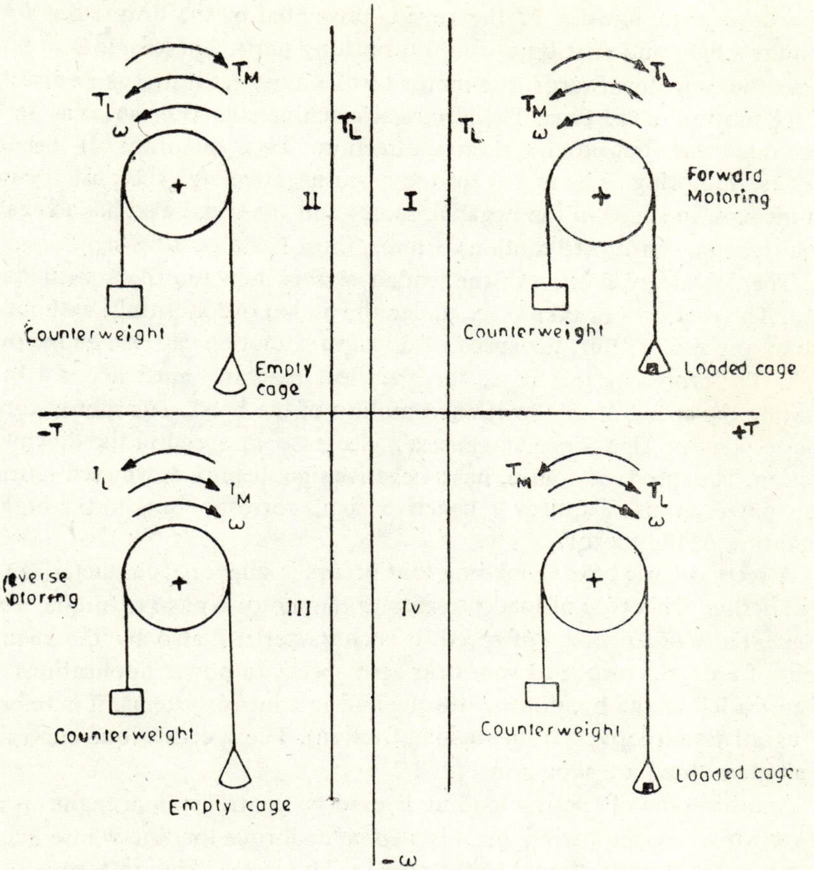

Fig. 2.1. Four quadrant operation of a motor driving a hoist load.

of rotation. Hence, to drive the loaded hoist up, the developed torque in the motor T_M must act in the same direction as the speed of rotation, i.e., T_M should be of positive sign. Since the speed is also of positive sign, being an upward motion, the power will also have a positive sign, i.e., the drive is said to be 'motoring'. Quadrant I is arbitrarily and conventionally, thus, designated as 'forward motoring quadrant'.

The hoisting up of the unloaded cage is represented in the second quadrant. Since the counterweight is heavier than the empty cage, the speed at which the hoist is moved upwards may reach a dangerously high value. In order to avoid this, the motor torque must act opposite to the direction of rotation, i.e., the motor should switch over to a braking or generator regime. Note that T_M will have a negative sign and speed still has a positive sign, giving power a negative sign, corresponding to the generator or braking operation.

The third quadrant represents the downward motion of the empty cage. The downward journey of the cage is prevented by the torque due to the counterweight and friction at the transmitting parts. Therefore, in order to move the cage downwards, the motor torque must act in the same direction as the motion of the cage. The electrical machine acts as a motor as in the first quadrant, but in the reverse direction. Thus quadrant III becomes 'reverse motoring'. The motor torque has a negative sign (because it causes an increase in speed in the negative sense) and the speed also has a negative sign (being a downward motion). Power, thus, has a positive sign.

The downward motion of the loaded cage is shown in the fourth quadrant. The motion can take place under the action of load itself, without the use of any motor. But, the speed of downward motion can be dangerously high. Therefore, in this case, the electrical machine must act as a brake limiting the speed of the downward motion of the hoist. The motor torque has a positive sign since it causes a decrease in speed in the downward motion. The speed, of course, has a negative sign, being a downward journey. The power, thus, acquires a negative sign, corresponding to the braking operation of the motor.

A second basic type of loading that occurs is the one characterized by dry friction. This type of load presents to the motor a passive torque, which is essentially independent of speed. It is characterized also by the requirement of an extra torque at very near zero speed. In power applications it is, often, called as the break away torque and in control systems, it is referred to as stiction (derived from sticking friction). The speed-torque curves for this type of load are shown in Fig. 2.2.

Another type of friction loading is used by control system engineers and is known as viscous friction. It is a force or torque loading whose magnitude is directly proportional to the speed. The viscous friction torque speed curves are illustrated in Fig. 2.3. Calendering machines, Eddy current brakes

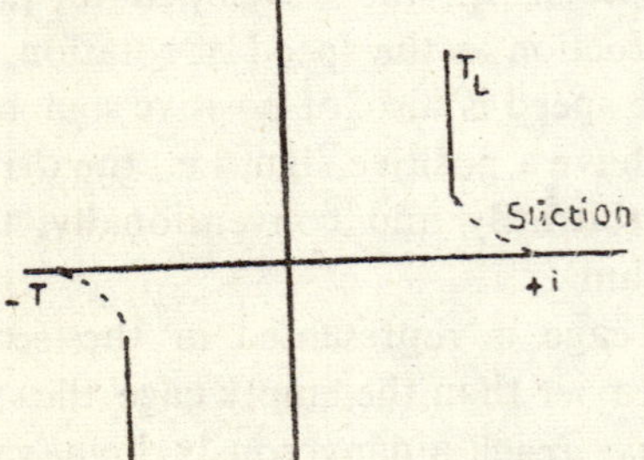

Fig. 2.2. Speed torque curve of dry friction load.

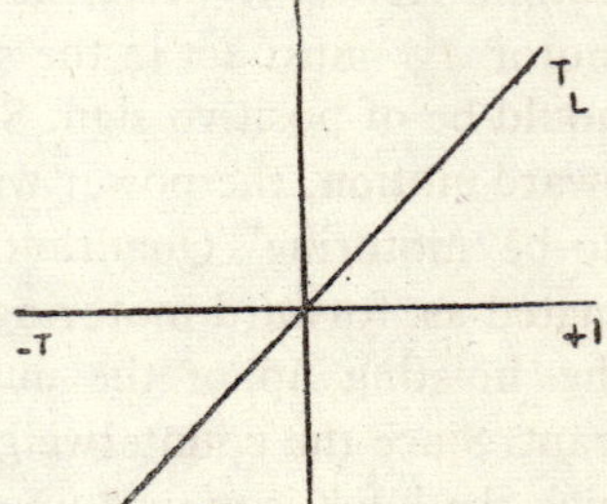

Fig. 2.3. Speed torque curves of viscous friction load.

and separately excited dc generators feeding fixed resistance loads have such speed-torque characteristics.

Yet another basic type of load torque is one whose magnitude is proportional to some power of the speed. Such a load is best illustrated by a fan or blower. The torque produced by the fan is directly proportional to the square of the speed throughout the range of usable fan speeds. The speed-torque curves for the fan type of load are presented in Fig. 2.4. Centrifugal pumps, propellers in ships or aeroplanes also have the same type of speed-torque characteristic.

Hyperbolic speed-torque characteristic (load torque being inversely proportional to speed or load power remaining constant), as shown in Fig. 2.5, is associated with certain type of lathes, boring machines, milling machines, steel mill coilers, etc.

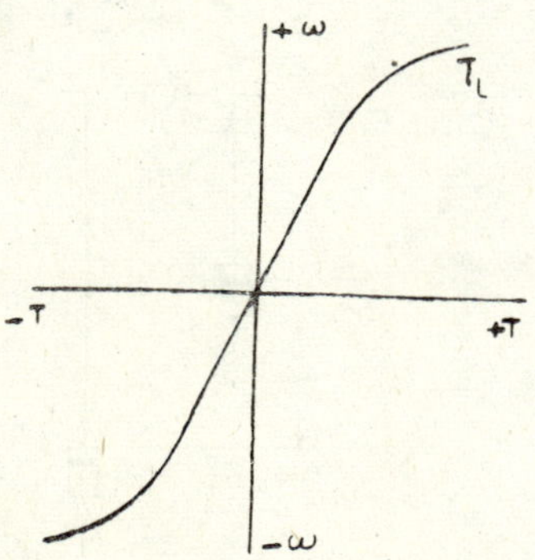

Fig. 2.4. Speed torque curve of a fan type load.

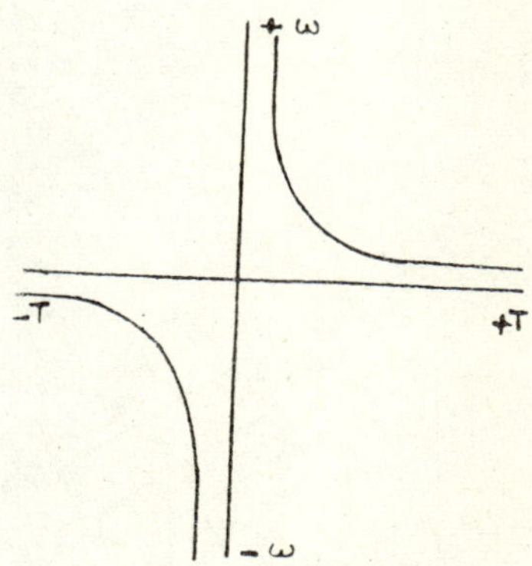

Fig. 2.5. Speed torque curve of a constant power load.

In general, the load torque in any specified application may consist of any of the above mentioned loads in varying proportions.

2.3 Load Torques that Depend on the Path or Position Taken by the Load During Motion

In the preceding section, we have been considering load torques which vary as a function of speed. However, load torques, that depend not only on speed but also on the nature of the path traced out by the load during its motion, are present both in hoisting mechanisms and transport systems. For instance, the resistance to motion of a train travelling upgradient or taking a turn depends on the magnitude of the gradient or the radius of curvature of the track respectively.

The force resisting the motion of a train travelling upgradient, as shown in Fig. 2.6 is given by

$$F_G = W \sin \alpha \cong W \tan \alpha \ (\alpha, \text{ being usually small})$$

$$= W \frac{G}{1000} \text{ kg}, \qquad (2.1)$$

where W = dead weight of the train or any other transport system, in kg, and G = gradient expressed as a rise in metres in a track distance of 1000 metres.

The tractive force required to overcome curve resistance is given by the empirical formula stated below:

$$F_c = \frac{700}{R} W \text{ kg}, \tag{2.2}$$

where R is the radius of curvature in metres.

In hoisting mechanisms in which tail ropes or balancing ropes are not used (Fig. 2.7) the load torque is not only due to the weight of the unloaded or the loaded cage but also due to that of the lifting ropes or cables.

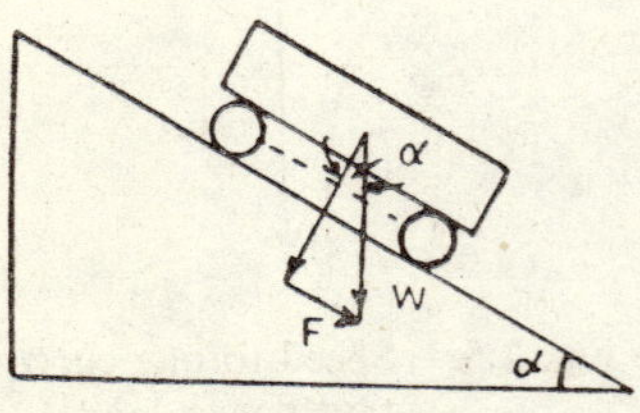

Fig. 2.6. Forces during the upgradient motion of a train.

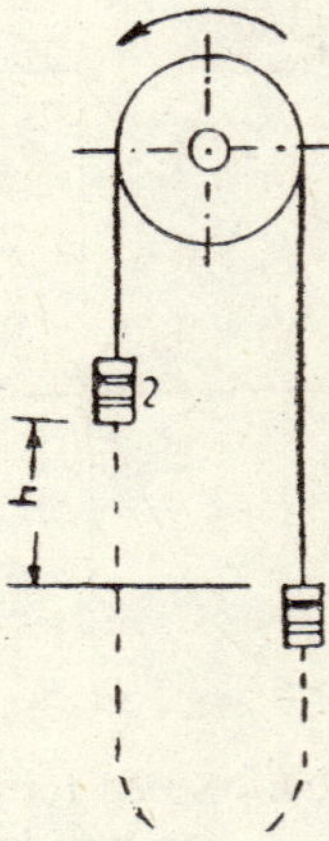

Fig. 2.7. Hoisting mechanism.

The latter depends on the position of the two cages. When cage 1 is at the bottom most position and is to be lifted upwards, the entire weight of the rope is also to be moved up. When both cages remain at the same height, the weight of the rope to be lifted up becomes zero, since the weight of the ropes on both sides balance each other. When cage 1 is at a higher position than cage 2, a portion of the weight of the rope acts in such a way as to aid the upward motion of cage 1. In fact, when cage 1 occupies the topmost position, the whole weight of the rope aids the upward movement.

The force that resists the upward motion of the load F_r due to the varying weight of the rope depending on the position of the load, is given as:

$$F_r = W_r \left(1 - \frac{2x}{h}\right) \text{ kg}, \tag{2.3}$$

where W_r = total weight of the rope, in kg,

x = height of the cage at any arbitrary position from the bottom most position in m, and

h = the desired maximum height to which the cage is to be moved upwards, in m.

Since, for very high values of 'h', the weight of the rope may be considerably greater than that of the load to be lifted upwards, the force F_r affects, to a large extent, the performance of the drive used in hoisting mechanisms. By using tail ropes, as shown by means of dotted lines in Fig. 2.7, the weight of the connecting rope can be balanced and more or less smooth movement of the cages can be ensured.

Another example of a load torque, which depends on path traced out during motion, is that of a planing machine. At a particular position of the moving table containing the workpiece, the load torque comes in the form of a sudden blow; in a different position, after the cutter has come out of the job, the magnitude of the load torque decreases sharply.

2.4 Load Torques That Vary with Angle of Displacement of the Shaft

In all machines, having crankshafts, for example, in reciprocating pumps and compressors, frame saws, weaving looms, rocking pumps used in petroleum industry etc., load torque is a function of the position of the crank, i.e., the angular displacement of the shaft or rotor of the motor. Load torque in drives used for steering ships also belongs to this category.

Figure 2.8 shows the approximate relationship between the load torque and angular displacement of the shaft 'θ' for a reciprocating compressor. It is of the form $T_L = f(\theta)$, where θ itself varies with time. For all such machines, the load torque T_L can be resolved into two components—one of constant magnitude T_{av} and the other a variable T'_L, which changes periodically in magnitude depending on the angular position of the shaft. Such load torque characteristics, can, for simplicity, be represented by

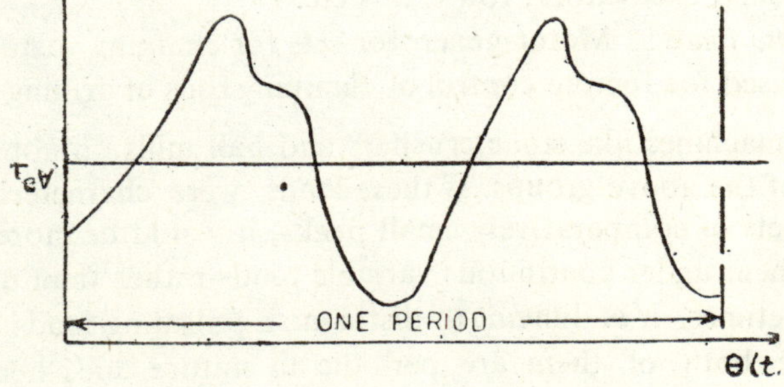

Fig. 2.8. Speed-torque characteristic of a reciprocating compressor.

Fourier series as a sum of oscillations of fundamental and harmonic frequencies, i.e.,

$$T'_{\mathrm{L}} = \sum_{r=0}^{m} T'_{\mathrm{L}r} \sin(r\theta + \phi_r) \tag{2.4}$$

$\theta = \omega t$, where ω represents the angular speed of the shaft of the motor driving the compressor.

During changes in speed, since only small deviations from a fixed value of speed ω_a occur, the angular displacement can be represented by $\theta = (\omega_a + \Delta\omega)\, t$. Then, the variable portion of the load torque may be expressed as

$$T'_{\mathrm{L}} = \sum_{0}^{m} T'_{\mathrm{L}r} \sin[(r\omega_a t + \phi_r) + r.\Delta\omega t] \tag{2.5}$$

The term $r.\Delta\omega t$ being of very small magnitude can be neglected. Thus, restricting to small deviations in angle from the equilibrium position, a load torque which varies with the angular displacement of the shaft can be transformed to one which varies periodically with respect to time.

2.5 Load Torques That Vary with Time

Of equal or perhaps greater importance in motor selection is the variation of load torque with time. This variation, in certain applications, can be periodic and repetitive, one cycle of variation being called a duty cycle. It is convenient to classify different types of loads under the following groups:

(a) *Continuous, constant loads*: Centrifugal pumps or fans operating for a long time under the same conditions; paper-making machines etc.

(b) *Continuous, variable loads*: Metal cutting lathes; hoisting winches; conveyors etc.

(c) *Pulsating loads*: Reciprocating pumps and compressors; frame saws, textile looms and, generally, all machines having crank shaft.

(d) *Impact loads*: Apparent, regular and repetitive load peaks or pulses which occur in rolling mills, presses, shearing machines, forging hammers etc. Drives for such machines are characterized by heavy flywheels.

(e) *Short time intermittent loads*: Almost all forms of cranes and hoisting mechanisms; excavators; roll trains etc.

(f) *Short time loads*: Motor-generotor sets for charging batteries; servomotors used for remote control of clamping rods of drilling machines.

Certain machines like stone crushers and ball mills do not strictly fall under any of the above groups. If these loads were characterized by frequent impacts of comparatively small peaks, it would be more appropriate to classify them under continuous variable loads rather than under impact loads. Sometimes, it is difficult to distinguish pulsating loads from impact loads, since both of them are periodic in nature and, hence, may be expressed as a sum of sinusoidal waves of different amplitude, frequency and phase.

One and the same machine can be represented by a load torque which varies either with speed or with time. For example, a fan load whose load torque is proportional to the square of the speed, is also a continuous,

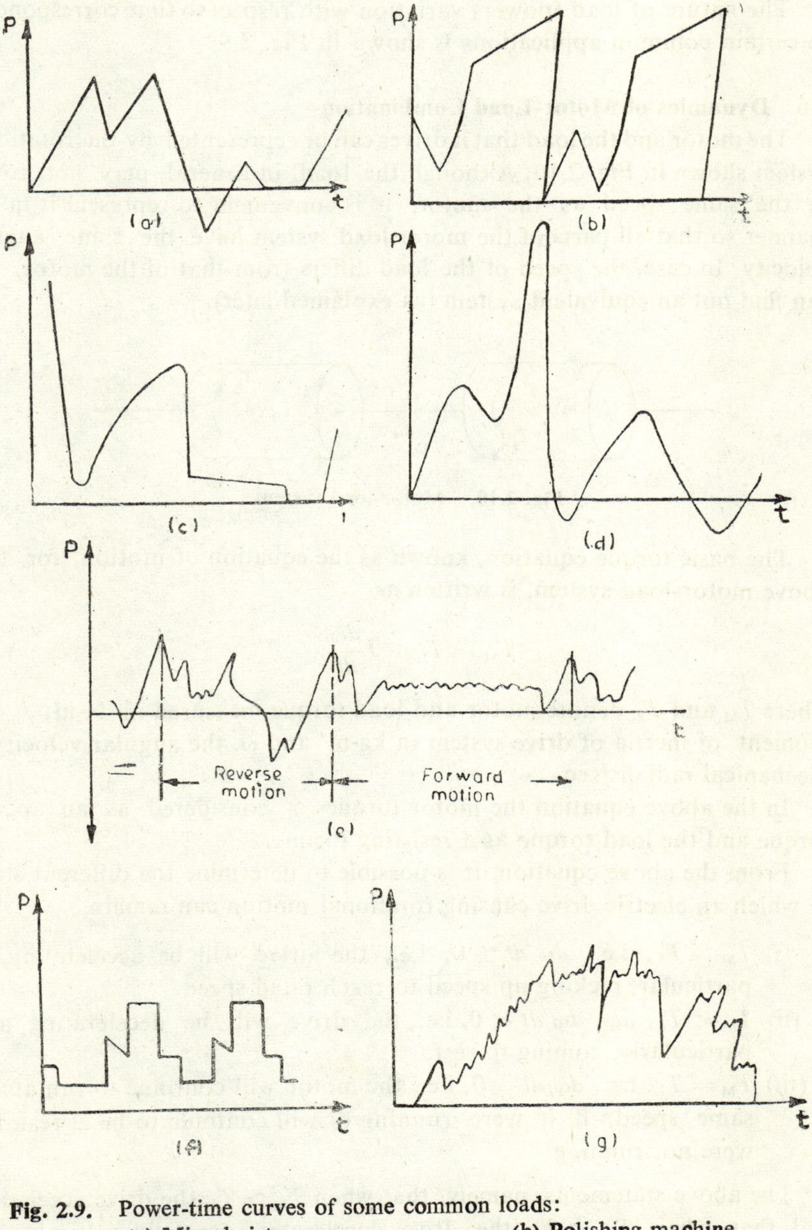

Fig. 2.9. Power-time curves of some common loads:
(a) Mine hoist (b) Polishing machine
(c) Shearing machine for cutting steel
(d) Textile loom (e) Planing machine
(f) Drilling machine (g) Grinding machine.

constant load. Load torque of a crane is independent of speed and also short time intermittent in nature. Rocking pumps for petroleum have a load which vary with angular position of the shaft, but can also be classified as a pulsating load.

The nature of load (power) variation with respect to time corresponding to certain common applications is shown in Fig. 2.9.

2.6 Dynamics of Motor-Load Combination

The motor and the load that it drives can be represented by the rotational system shown in Fig. 2.10. Although the load, in general, may not rotate at the same speed as the motor, it is convenient to represent it in this manner so that all parts of the motor-load system have the same angular velocity. In case, the speed of the load differs from that of the motor, one can find out an equivalent system (as explained later).

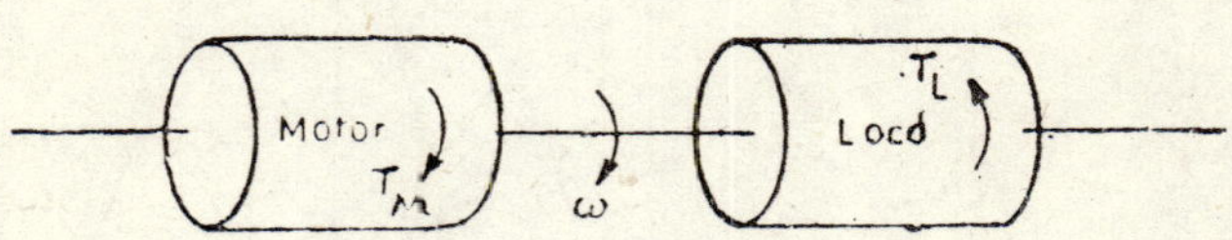

Fig. 2.10. Motor-load system.

The basic torque equation, known as the equation of motion, for the above motor-load system, is written as

$$T_M = T_L + J\frac{d\omega}{dt} \tag{2.6}$$

where T_M and T_L denote motor and load torque measured in N-m; J, the moment of inertia of drive system in kg-m^2 and ω, the angular velocity in mechanical radians/sec.

In the above equation the motor torque is considered as an applied torque and the load torque as a resisting torque.

From the above equation, it is possible to determine the different states at which an electric drive causing rotational motion can remain.

(i) $T_M > T_L$, i.e., $d\omega/dt > 0$, i.e., the drive will be accelerating, in particular, picking up speed to reach rated speed.

(ii) $T_M < T_L$, i.e., $d\omega/dt < 0$, i.e., the drive will be decelerating and, particularly, coming to rest.

(iii) $T_M = T_L$, i.e., $d\omega/dt = 0$, i.e., the motor will continue to run at the same speed, if it were running or will continue to be at rest, if it were not running.

The above statements, namely, that when $T_M > T_L$ the drive accelerates and that when $T_M < T_L$ the drive decelerates, are valid only when T_L happens to be a passive load. The reverse may occur with active loads. For example, if we were to switch on the motor for hoisting up a winch, while it is coming down on its own weight, until the direction of rotation changes, deceleration of the drive and not acceleration takes place, when $T_M > T_L$. In case $T_M < T_L$ in the above situation when the motor has been switched on for moving the winch up, the load will continue to come down and the motor will accelerate instead of decelerating.

The term $J\,d\omega/dt$ which represents the inertia torque, is also known as dynamic torque, since it is present only during transient conditions, i.e., when the speed of the drive varies. During acceleration of the drive, the inertia torque is directed against motion, but during braking it maintains the motion of the drive. Thus, inertia torque is determined both in magnitude and sign, as the algebraic sum of the motor and load torques.

In view of the above, the signs for T_M and T_L in Eqn. (2.6) correspond to motoring operation of the driving machine and to passive load torque or to a braking torque caused by active loads, respectively. The equation of motion can, in general, be written as:

$$\pm T_M = \pm T_L + \frac{J\,d\omega}{dt} \tag{2.7}$$

The signs to be associated with T_M and T_L in Eqn. (2.7) depend, as indicated earlier, on the regime of operation of the driving motor and the nature of load torque. The equation of motion enables us to determine the variation of torque, current and speed with respect to time, during transient operation of the drive.

2.6.1 Equivalent System

Seldom is a motor shaft directly coupled to load shafts. In general, the different loads connected to the motor will have different speed requirements. Speed changing mechanisms such as gears, V-belts, etc., will be used to obtain different speeds. Since the ultimate objective is to select a motor suitable for the application, it is desirable to refer all mechanical quantities such as load torque, inertia torque, etc., to one single axis of rotation, conveniently, the output shaft of the motor. The principle of conservation of energy will be used for this purpose.

2.6.2 Determination of Referred Load Torque

Let the speed of the motor shaft be ω_M and that of the load be ω_L.

Equating power, we have

$$T_L \cdot \omega_L \frac{1}{\eta} = T'_L\, \omega_M$$

i.e.,

$$T'_L = T_L \cdot \frac{\omega_L}{\omega_M} \times \frac{1}{\eta} = \frac{T_L}{i\eta}, \tag{2.8}$$

where T_L = load torque

T'_L = load torque referred to the motor shaft,

$i = \frac{\omega_M}{\omega_L}$ = speed transmission ratio (gear ratio), and

η = efficiency of transmission.

When there are several stages in transmission between the driving motor and the driven machine, as shown in Fig. 2.11 with gear ratios $i_1, i_2, ..., i_n$ and the respective efficiencies $\eta_1, \eta_2, ..., \eta_n$, the load torque referred to the motor shaft is given as

$$T'_L = T_L \times \frac{1}{i_1 \cdot i_2 \cdots i_n} \times \frac{1}{\eta_1 \cdot \eta_2 \cdots \eta_n} \tag{2.9}$$

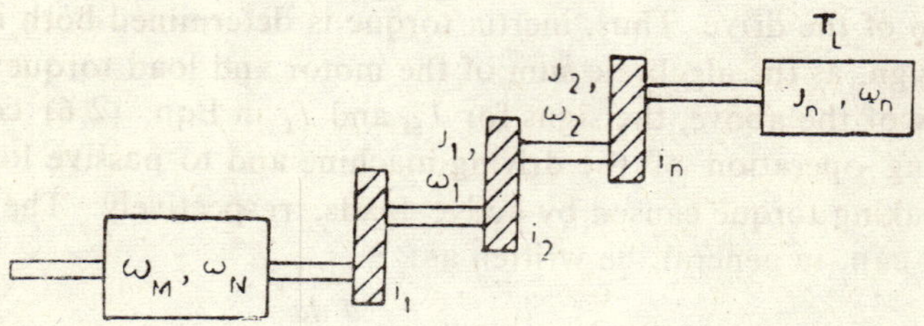

Fig. 2.11. Motor-Load system with gears.

2.6.3 *Determination of Referred Moment of Inertia*

Moments of inertia are referred to a given shaft on the basis that the total amount of kinetic energy stored in the moving parts and referred to the given shaft remains unchanged. With the rotating parts having moments of inertia $J_M, J_1, J_2, ..., J_n$ and angular speeds $\omega_M, \omega_1, \omega_2, ..., \omega_n$ (Fig. 2.11), the equivalent moment of inertia referred to the shaft may be found as follows:

$$\frac{J'\omega_M^2}{2} = J_M \frac{\omega_M^2}{2} + J_1 \frac{\omega_1^2}{2} + J_2 \frac{\omega_2^2}{2} + \ldots + J_n \frac{\omega_n^2}{2},$$

where J', the moment of inertia referred to the motor shaft

$$= J_M + J_1 \left(\frac{\omega_1}{\omega_M}\right)^2 + J_2 \left(\frac{\omega_2}{\omega_M}\right)^2 + \ldots + J_n \left(\frac{\omega_n}{\omega_M}\right)^2$$

$$= J_M + \frac{J_1}{i_1^2} + \frac{J_2}{(i_1 i_2)^2} + \ldots + \frac{J_n}{(i_1 i_2 \ldots i_n)^2} \tag{2.10}$$

2.6.4 *Referring Forces and Masses Having Translational Motion to a Rotating Shaft*

In many machines, for example, in hoists, cranes and shaping machines, some of the moving parts rotate while others go through a translational motion.

Figure 2.12 shows the forces and torques acting on a hoist drive. If the moving mass has a velocity of v m/sec and the motor shaft has an angular velocity ω_M rad/sec,

$$F_r \times v \frac{1}{\eta_l} = T'_L \, \omega_M,$$

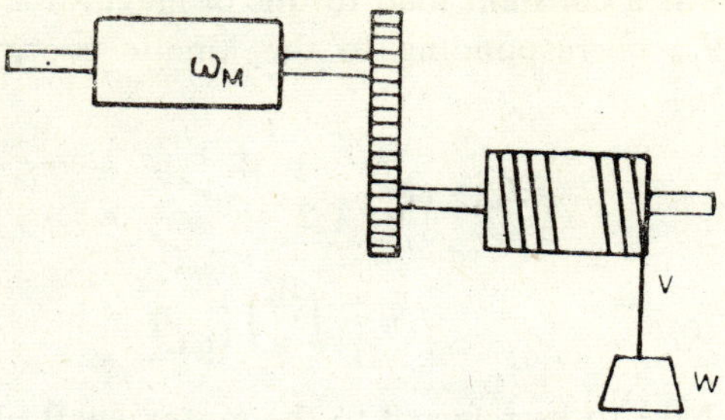

Fig. 2.12. Motor-hoist load system.

where F_r is the resisting force developed by the load due to the gravitational pull of the moving weight W and η, the efficiency of transmission.

Referred load torque $T'_L = \frac{F_r . v}{\omega_M . \eta}$ (2.11)

A mass having translational motion is referred to a rotating one on the basis of constant kinetic energy

$$\frac{mv^2}{2} = J' \frac{\omega_M^2}{2}$$

The M.I. referred to the motor shaft

$$J' = m(v/\omega_m)^2 = W/g.(v/\omega_M)^2 \quad (2.12)$$

Example 2-1. Determine the equation of motion of the drive system consisting of a motor, a single gear train, an inertia torque, a hoist load, a dry friction load, a viscous friction load and a fan load as shown in Fig. 2.13.

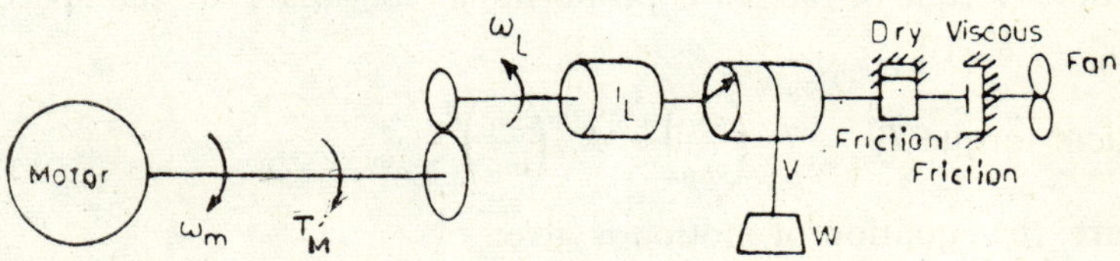

Fig. 2.13. Drive system for Example 2-1.

Solution: First, all mechanical quantities have to be referred to the motor shaft. Let us denote all the referred quantities as primed ones.

INERTIA TORQUE: Equating kinetic energy, we have

$$\frac{1}{2} J_L . \omega_L^2 = \frac{1}{2} J'_L . \omega_m^2$$

Hence, $$J'_L = \left(\frac{\omega_L}{\omega_m}\right)^2$$

HOIST LOAD: There is a constant load torque of magnitude $T_{Lh} = W.r$ and an inertia torque J_{Lh} corresponding to the kinetic energy stored by the weight W.

$$\frac{1}{2} J_{Lh} \times \omega_L^2 = \frac{1}{2} \cdot \frac{W}{g} . v^2$$

Hence,

$$J_{Lh} = \left(\frac{W}{g}\right)\left(\frac{v}{\omega L}\right)^2$$

Both T_{Lh} and J_{Lh} have to be referred to the motor shaft. Equating power, we get

$$T_{Lh} \times \omega_L = T'_{Lh} \times \omega_m$$

so

$$T'_{Lh} = T_{Lh}\left(\frac{\omega_L}{\omega_m}\right)$$

Equivalent inertia torque $J'_{Lh} = J_{Lh}\left(\frac{\omega_L}{\omega_m}\right)^2 = \frac{W}{g}\left(\frac{v}{\omega_m}\right)^2$

DRY FRICTION: It is a constant load torque of magnitude, say T_d.

Equivalent torque $T'_d = T_d\left(\frac{\omega_L}{\omega_m}\right)$

VISCOUS FRICTION: This is a load torque proportional to speed, i.e., $T_v \propto \omega_L = K_1 \cdot \omega_L$

Equivalent torque $T'_v = T_v\left(\frac{\omega_L}{\omega_m}\right) = K_1\left(\frac{\omega_L}{\omega_m}\right)^2 \cdot \omega_m$

FAN: Load torque of the fan is proportional to square of the speed, i.e., $T_f \propto \omega_L^2 = K_2 \cdot \omega_L^2$

Equivalent torque $T'_f = T_f\left(\frac{\omega_L}{\omega_m}\right) = K_2\left(\frac{\omega_L}{\omega_m}\right)^3 \cdot \omega_m^2$

Therefore, the equation of motion is given by

$$T_M = J_L \cdot C^2 + \left(\frac{W}{g}\right)\left(\frac{v}{\omega_m}\right)^2 + W \cdot r \cdot C + T_d \cdot C + K_1 C^2 \cdot \omega_m + K_2 C^3 \cdot \omega_m^2,$$

where

$$C = \left(\frac{\omega_L}{\omega_m}\right).$$

2.6.5 *Referring Torques and Masses which Undergo Translational Motion at Variable Speeds*

In certain types of machines which convert the rotational motion into a translational one with the help of a crankshaft, the speed and acceleration of the moving masses vary both in magnitude and in sign during each revolution of the crankshaft. The kinetic energy stored in these masses will change from zero to a maximum. The moment of inertia of these masses referred to the crankshaft is

$$J' = \frac{mv^2}{\omega^2}, \tag{2.13}$$

where m is the mass of the given body which undergoes the translational motion with a velocity v and ω, the angular velocity of the crankshaft.

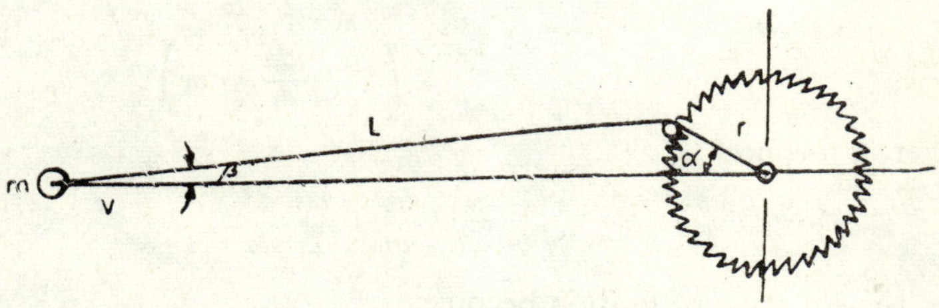

Fig. 2.14. Forces in a crankshaft mechanism.

Using Fig. 2.14, the relationship between the linear velocity of the piston v and the angular speed of the crankshaft can be found as:

$$v = \frac{r\,\omega \sin(\alpha + \beta)}{\cos\beta}, \tag{2.14}$$

where $\beta = \sin^{-1}\left(\frac{r}{L}\sin\alpha\right)$. The value of α depends on the position of the crank pin. Substituting v from Eqn. (2.14), the moment of inertia referred to crankshaft

$$J' = \frac{mr^2 \sin^2(\alpha + \beta)}{\cos^2\beta} \tag{2.15}$$

The total moment of inertia will be the sum of the moment of inertia J' and that of all other moving parts referred to the crankshaft. If there is some intermediate stage of transmission between the crankshaft and the driving motor, the moments of inertia referred to the crankshaft should then be referred to the motor shaft, using Eqn. (2.11).

Hence, the load torque referred to the motor shaft

$$T'_L = \frac{F \cdot r \sin(\alpha + \beta)}{\eta i \cos\beta}, \tag{2.16}$$

where F is the resisting force offered by that part, which undergoes translational motion, i the gear ratio ω_M/ω, if any and η the efficiency of transmission.

In mechanisms using crankshafts, the moment of inertia varies as a function of α and hence the equation of motion has a more complex form.

Kinetic energy stored in the crankshaft

$$(\text{K.E.})_c = \frac{J\omega^2}{2} \tag{2.17}$$

The dynamic power

$$P_{\text{dyn}} = \frac{d(\text{K.E.})_c}{dt}$$

$$= J\omega\frac{d\omega}{dt} + \frac{\omega^2}{2}\frac{dJ}{d\alpha}\cdot\frac{d\alpha}{dt}$$

$$= J\omega\frac{d\omega}{dt} + \frac{\omega^3}{2}\cdot\frac{dJ}{d\alpha} \tag{2.18}$$

$$\left(\because \quad \frac{d\alpha}{dt} = \omega\right)$$

The inertia torque T_{dyn}

$$= P_{\text{dyn}}/\omega$$

$$= J\frac{d\omega}{dt} + \frac{\omega^2}{2}\cdot\frac{dJ}{d\alpha} \tag{2.19}$$

Hence, the equation of motion becomes

$$T_M - T_L = J\cdot\frac{d\omega}{dt} + \frac{\omega^2}{2}\cdot\frac{dJ}{d\alpha} \tag{2.20}$$

The additional term, that has appeared is obviously due to the variable moment of inertia.

2.7 Determination of Moment of Inertia

While analysing the transient performance of a drive, it is necessary to know the value of moment of inertia of the rotating parts. It can either be determined analytically by using design data of the moving parts or experimentally by the following methods.

Retardation Test: The most common and simple method of determining J is by performing a test known as retardation test on the drive. During this test, the motor is run upto a speed slightly higher than the normal and the supply to it is cut off. The power input to the motor before switching off the supply is noted. An oscillographic record of the speed of the motor at different instants of time after the supply has been cut off is made.

Now, if the source of energy to a rotational system is cut off, it will continue to rotate due to the initial kinetic energy stored in the system. But as this energy is used up to supply the rotational losses in the system, it slows down and gradually stops. The power consumed in overcoming the rotational losses is given by

$$P = \text{rate of change of kinetic energy}$$

$$= \frac{d}{dt}\left(\frac{1}{2}J\omega^2\right)$$

$$= J \times \frac{4\pi^2}{3600} N\cdot\frac{dN}{dt}, \tag{2.21}$$

where N is in rpm.

From the test results obtained by performing the retardation test, dN/dt at normal rated speed N can be found graphically using the oscillogram of speed vs. time. As a first approximation, the measured input power to the motor may be taken as P. Now, using Eqn. (2.21), J can be calculated.

Another method, which is more accurate than the one described earlier, uses the speed vs. time curve obtained during the retardation test, but determines the rotational losses more accurately. By means of additional experiment, we must get the plot of rotational losses against speed for the motor under consideration. Both the curves are plotted in the same graph as shown in Fig. 2.15. Using the above two curves, it is possible to get a

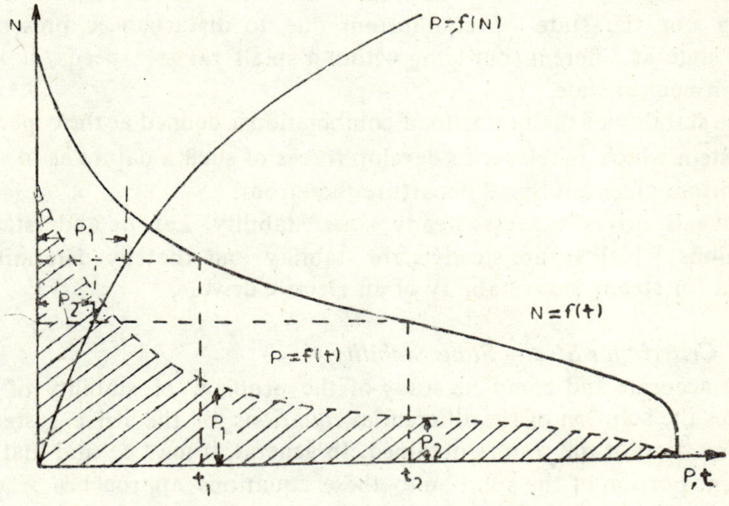

Fig. 2.15. Graphical method for determination of J.

relationship between rotational losses and time. Figure 2.15 also shows how $P = f(t)$ can be plotted. The area under this curve is nothing but the kinetic energy used up by the rotating masses while retardation took place, i.e., $J\,\omega^2/2$ = Area A.

$$\therefore \qquad J = A/N^2 \times 1800/\pi^2 \qquad (2.22)$$

2.8 Steady State Stability of an Electric Drive

It is quite important to investigate the conditions of stable operation of an electric drive, when its equilibrium state is disturbed. In practice, it is necessary to consider two types of disturbance:

(1) Changes from the state of equilibrium take place slowly and the effect of either the inertia of the rotating masses or that of inductance is insignificant.

(2) Sudden and fast changes from the equilibrium state, as a result of which the effect of neither the inertia nor the inductane can be ignored.

Study of stability under conditions enunciated for the first type of disturbance relate to the field of steady state stability, while for the second type of disturbance pertain to the field of dynamic or transient stability.

The drive is said to be in equilibrium, if at a given speed the torque developed by the motor is exactly equal to the load torque. However, this state of equilibrium may be stable, unstable or neutral, as observed in other physical processes. If the drive comes out of the state of equilibrium under the influence of any unexpected external (change in the amount of energy fed to the motor) or internal (change in load) factors, in the case of stable equilibrium, the drive again returns to the steady state, but in the case of unstable equilibrium the speed of the drive either increases tremendously or decreases and comes to rest. When the drive, after coming out the state of equilibrium due to disturbances, preserves its steady state at different (but lying within a small range) speeds, it is said to be in neutral state.

The stability of the motor-load combination is defined as the capacity of the system which enables it to develop forces of such a nature as to restore equilibrium after any small departure therefrom.

Not all drives possess steady state stability. Let us understand the conditions, which ensure steady state stability and thereby determine the criteria for steady state stability of an electric drive.

2.8.1 *Criteria for Steady State Stability*

An accurate and complete study of the problem of stability of drives requires the solution of the differential equations of the drive system and interpretation of the results obtained. In general, it may be said that, if the transient portion of the solution to these equations approaches zero with increasing time, the effects of any disturbance die away, and the system is stable.

However, in the absence of any such complete analysis, considerable insight to the problem may be gained by a study of the steady state speed torque curves, of both motor and load, and assuming that all departures from equilibrium will be along these curves.

Let the equilibrium values of the torques and speed be denoted by T_M, T_L and ω and let small deviations be denoted by ΔT_M, ΔT_L and $\Delta\omega$. After a small displacement from the equilibrium, the torque equation becomes

$$J \cdot \frac{d\omega}{dt} + J \cdot \frac{d(\Delta\omega)}{dt} + T_L + \Delta T_L - T_M - \Delta T_M = 0 \qquad (2.23)$$

But
$$J \cdot \frac{d\omega}{dt} + T_L - T_M = 0 \qquad (2.24)$$

Hence,
$$J \cdot \frac{d(\Delta\omega)}{dt} + \Delta T_L - \Delta T_M = 0 \qquad (2.25)$$

The above equation is to be interpreted as the torque equation in which all quantities are expressed in terms of their deviations from equilibrium values. If we assume that these increments are so small that they may be expressed as linear functions of the change in speed, then

$$\Delta T_M = \frac{dT_M}{d\omega} \cdot \Delta\omega, \quad (2.26)$$

$$\Delta T_L = \frac{dT_L}{d\omega} \cdot \Delta\omega, \quad (2.27)$$

where $dT_M/d\omega$ and $dT_L/d\omega$ indicate derivatives at the point of equilibrium. Substituting these relations in Eqn. (2.25) and rearranging, we have

$$J \cdot d\frac{(\Delta\omega)}{dt} + \left[\frac{dT_L}{d\omega} - \frac{dT_M}{d\omega}\right] \Delta\omega = 0, \quad (2.28)$$

whose solution is

$$\Delta\omega = (\Delta\omega)_0 e^{-1/J\left[\frac{dT_L}{d\omega} - \frac{dT_M}{d\omega}\right] t} \quad (2.29)$$

The quantity $(\Delta\omega)_0$ is the initial value of the deviation in speed. In order that the system be stable, it is necessary that the exponent be negative, so that the speed increment will disappear with time. Under such conditions, the system will return to its equilibrium speed. Whereas if the exponent is greater than zero the speed deviation will increase with time, and the system will move away from equilibrium. If the exponent is exactly equal to zero, the equation is insufficient to discuss about stability. The exponent will always be negative if

$$\frac{dT_L}{d\omega} - \frac{dT_M}{d\omega} > 0 \quad (2.30)$$

This is equivalent to saying that for a decrease in speed the motor torque must exceed the load torque and for an increase in speed the motor torque must be less than the load torque.

This relationship is illustrated in Fig. 2.16 in which the load torque T_L' results in a stable operating point, and the load torque curve T_L'' results in an unstable situation.

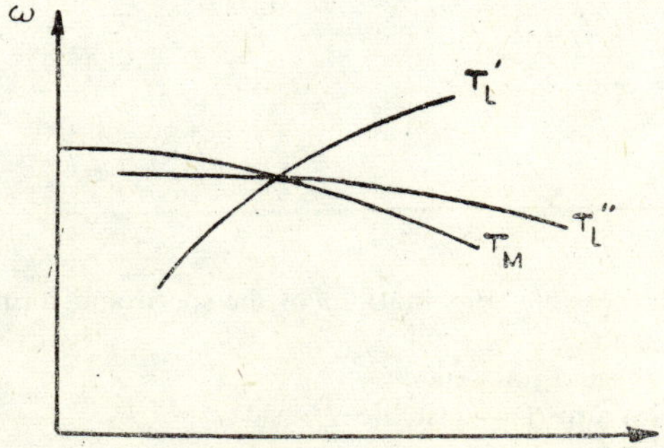

Fig. 2.16. Speed-torque curves of motor-load combination.

The stability of the operating point is most easily evaluated by assuming a small change in speed away from equilibrium. With this speed increment assumed, the relative values of motor and load torques will determine whether the speed will return to its previous value. That is, if an increase in speed brings about a greater increase in load torque than motor torque, the speed will tend to decrease and return to its original value, which is then a stable point. For the converse case, the speed continues to increase, and the system is obviously unstable.

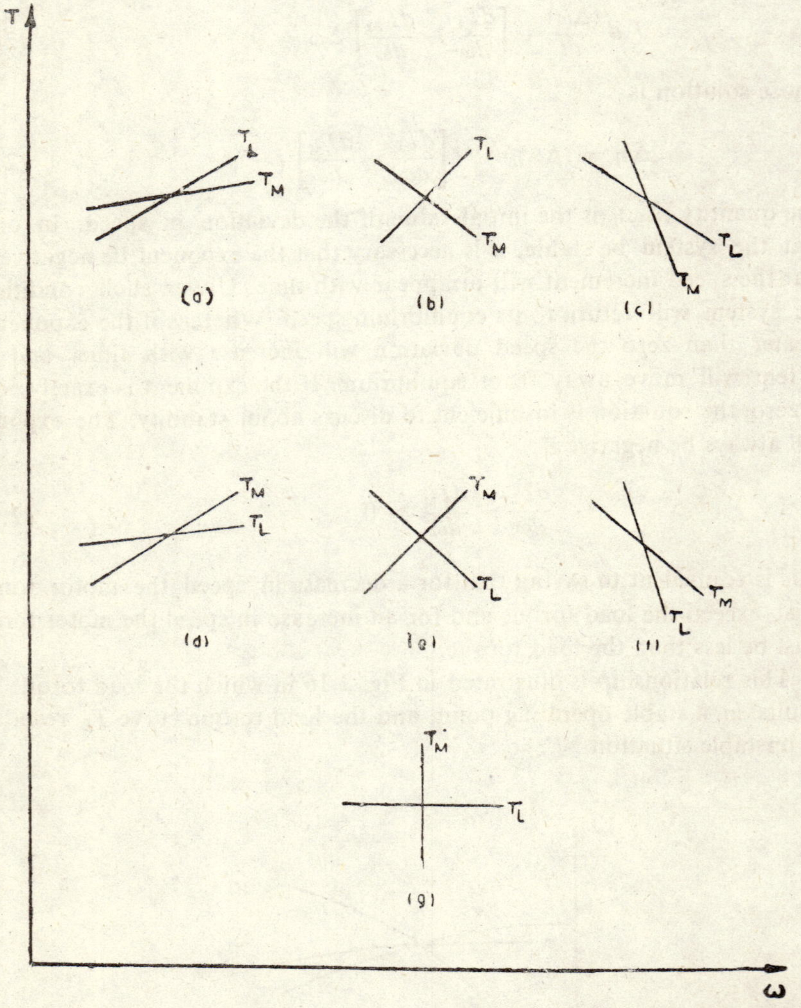

Fig. 2.17. Seven possible combinations of the speed and torque curves of motor and load:
(a), (b) and (c)—stable
(d), (e) and (f)—unstable
(g)—indeterminate.

It should be emphasized that this criterion for stability is an approximation in the sense that it assumes that the steady state speed-torque curves are traversed when small disturbances cause a departure in speed from some operating point. This assumption may or may not be valid, as there are many instances in which the transient speed-torque relations are quite different from the steady state ones. Despite this fact, the above criterion is quite useful.

Figure 2.17 shows seven possible combinations of motor and load characteristics and indicates which of them represent stable or unstable operation of the electric drive.

Example 2-2. A motor having a suitable control circuit develops a torque given by the relationship $T_M = a\omega + b$, where a and b are positive constants. This motor is used to drive a load whose torque is expressed as $T_L = c\omega^2 + d$, where c and d are some other positive constants. The total inertia of the rotating masses is J.

(a) Determine the relations amongst the constants a, b, c and d in order that the motor can start together with the load and have an equilibrium operating speed ?
(b) Calculate the equilibrium operating speed ?
(c) Will the drive be stable at this speed ?
(d) Determine the initial acceleration of the drive ?
(e) Determine the maximum acceleration of the drive ?

Solution: (a) At $\omega = 0$, $T_M = b$ and $T_L = d$.
Hence the motor can start with the load only if $b > d$.
$T_M = T_L$ at the equilibrium speed

i.e., $a\omega + b = c\omega^2 + d$

i.e., $c\omega^2 - a\omega - (b - d) = 0$

$$\text{Hence } \omega = \frac{a \pm \sqrt{a^2 + 4c(b-d)}}{2c}$$

In order that ω is finite $a^2 + 4c(b - d) > 0$, which is true.

$+$sign before the radical will give a positive ω as long as

$\sqrt{a^2 + 4c(b-d)} > 0$

$-$sign before the radical will give a positive ω only if

$$\frac{a}{2c} > \frac{\sqrt{a^2 + 4c(b-d)}}{2c}$$

i.e., $a^2 > a^2 + 4c(b - d)$

i.e., $4c\,(b - d) < 0$

i.e., $c < 0$, which is not true, since c is given to be a positive constant.

Hence the $+$sign before the radical only will give a positive finite equilibrium speed

if, $\sqrt{a^2 + 4c(b-d)} > 0$.

(b) Equilibrium speed $\omega = \dfrac{a + \sqrt{a^2 + 4c(b - d)}}{2c}$

(c) $\dfrac{dT_L}{d\omega} = 2c\omega$ and $\dfrac{dT_M}{d\omega} = a$

If the equilibrium speed has to be stable

$$\frac{dT_L}{d\omega} > \frac{dT_M}{d\omega} \qquad \text{i.e., } 2c\omega > a$$

From the answer to (b), we have

$2c\omega = a + \sqrt{a^2 + 4c(b - d)}$ which will be always $> a$.

Hence, the equilibrium operating speed determined earlier is a stable point of operation of drive.

(d) Accelerating torque $J\dfrac{d\omega}{dt} = T_M - T_L$

Initially $T_M = b$ and $T_L = d$

Therefore, initial acceleration $= \dfrac{b - d}{J}$.

(e) Accelerating torque $J\dfrac{d\omega}{dt} = T_M - T_L$

$$= a\omega - c\omega^2 + b - d$$

Therefore, acceleration $A = \dfrac{d\omega}{dt} = \dfrac{a\omega - c\omega^2 + b - d}{J}$

This will be maximum at a speed when

$$\frac{dA}{d\omega} = 0$$

i.e.,

$$\frac{a - 2c\omega}{J} = 0$$

i.e.,

$$\omega = \frac{a}{2c}$$

Substituting this speed at which the acceleration is maximum, in the general expression for acceleration, we get

$$A_{max} = \frac{a^2/2c - a^2/4c + b - d}{J}$$

$$= \frac{a^2 + 4c(b - d)}{4cJ}$$

2.9 Transient Stability of an Electric Drive

2.9.1 Concept of Transient Stability

While analysing the steady state stability of an electric drive, only its initial and final conditions are examined on the basis of the speed torque characteristics of the motor and load, without paying any attention to the inertia torques and to the time taken to change over from the initial condition to the final one. The nature of the motion of the drive during this period is also not considered. The essence of the study of transient stability of an electric drive or, in general, any moving system is to take into account the influence of the abovementioned factors. Such a study enables us to estimate more accurately the performance of the drive with respect to its stability of motion.

With slow and steady changes in load, the driving motor may be loaded upto its maximum capacity of torque or power. For example, an induction motor can be loaded upto its pull out torque, determined after taking into account any fall in voltage of the supply lines.

In case of transient processes, which takes place quite fast, the pull out torque of the motor no longer will be the permissible limit of load torque, since the inertia torque due to the kinetic energy of the rotating masses also would come into play. The inertia torque aids the motor torque, while the speed decreases and opposes the motor torque, while the speed increases. Thus, the equality between load torque and the motor torque, which forms the basis for the determination of steady state stability limit, no longer holds good during transient processes.

A study of the transient stability of an electric drive, both during its design and operation, enables us to use the equipment more rationally. For example, in case of drives having flywheels, the size of the latter can be decreased, if the design were to be done on the basis of transient stability limit. However, in practice, this is done only partially, allowing the transient stability limit to be used as a reserve capacity, when sudden changes in load occur.

2.9.2 Transient Stability of a Synchronous Motor

Let us consider a drive system consisting of a synchronous motor driving a constant torque load. The speed torque characteristics of the motor and load are shown in Fig. 2.17(g). From the figure, it is obvious that $\frac{dT_L}{d\omega} = 0$ and $\frac{dT_M}{d\omega} = \pm\infty$ (infinity). Hence, by applying the criterion of steady state stability, it can be seen that the operation of the drive system under consideration will be stable if $\left(\frac{dT_L}{d\omega} - \frac{dT_M}{d\omega}\right)$ is equal to $+\infty$ and unstable if it is equal to $-\infty$. That is, for the given drive system, the criterion of steady state stability does not give any useful information. Under such circumstances, to investigate stability using only the steady state characteristics of the motor is quite insufficient. It becomes necessary to study the essence of the processes involved during transition of synchronous motor from one operating condition to the other. In other words, it is imperative to investigate the transient stability of such a drive system.

If a large load is suddenly applied to the shaft of a synchronous motor, the motor must slow down momentarily at least to have a larger value of the torque angle required to supply the added load. In fact, until the new angle is attained, a considerable portion of the energy supplied to load comes from the stored energy in the rotating mass as it reduces its speed. As the required value of torque angle is reached, the rotor does not attain its equilibrium since the rotor speed is less than the synchronous speed. The torque angle, hence, must increase further in order to allow replenishment of the deficit of stored energy in the rotating mass. This process involves a series of oscillations of the rotor about its final position even when equilibrium is finally restored.

Similar oscillations accompanied with torque and current pulsations occur in synchronous motors driving loads whose torque requirements vary cyclically at a fairly rapid frequency, like in motors driving reciprocating air or ammonia compressors. If the natural frequency of mechanical oscillation of the rotor of the synchronous motor becomes equal or close to the frequency of a significant harmonic of the load cycle variations, oscillations of very high magnitude are produced. Exact description of such processes can only be given in terms of the related electromechanical differential equation and information about the restoration of equilibrium about the rotor can be observed only by solving the equation.

The equation of motion, in terms of power, can be written as,

$$P_M = P_{dyn} + P_L \tag{2.31}$$

where P_M, P_{dyn} and P_L denote the electromagnetic power developed by the motor, dynamic power and load power at the shaft, respectively.

The dynamic power is determined from the angular acceleration. The angular position of the shaft at any instant is taken as the electrical angle δ between a point on it and a reference which is rotating at synchronous speed. Often the angle δ is assumed to be the same as the torque or power angle. With a sudden application of load, since the rotor slows down, the angular acceleration will be negative and hence the dynamic power will be given by

$$P_{dyn} = -P_j \frac{d^2\delta}{dt^2}, \tag{2.32}$$

where

$$P_j = J.\omega.\frac{2}{\text{Poles}} \tag{2.33}$$

The electromagnetic power P_M usually has two components: (i) damping power, which is assumed to vary linearly with the departure $d\delta/dt$ from synchronous speed and (ii) synchronous power produced by synchronous motor action, which is a function of load angle δ.

Thus, the electromechanical equation becomes

$$P_j \frac{d^2\delta}{dt^2} + P_d.\frac{d\delta}{dt} + P(\delta) = P_L, \tag{2.34}$$

where P_d is the damping power per unit departure in speed.

Neglecting damping and assuming a cylindrical rotor synchronous machine, Eqn. (2.34) becomes

$$P_j \frac{d^2\delta}{dt^2} + P_m \sin \delta = P_L, \tag{2.35}$$

where $P_m = \frac{VE}{X_s}$ in which V, E and X_s denote the applied voltage, emf due to excitation and synchronous reactance of the motor.

From Eqn. (2.35), we get

$$\frac{d^2\delta}{dt^2} = \frac{P_L - P_m \sin \delta}{P_j}$$

Multiplying both sides by $\frac{d\delta}{dt}$, we have

$$\frac{d^2\delta}{dt^2}\left(\frac{d\delta}{dt}\right) = \left(\frac{P_L - P_m \sin \delta}{P_j}\right)\frac{d\delta}{dt}$$

so,
$$\frac{1}{2} \cdot \frac{d}{dt}\left(\frac{d\delta}{dt}\right)^2 = \left(\frac{P_L - P_m \sin \delta}{P_j}\right)\frac{d\delta}{dt}$$

so,
$$\frac{d\delta}{dt} = \sqrt{\int_{\delta_0}^{\delta} \frac{2(P_L - P_m \sin \delta)}{P_j} . d\delta},$$

where δ_0 is the load angle before the disturbance, i.e., at time $t = 0$. Also since the motor was running at synchronous speed at time $t = 0$, $\frac{d\delta}{dt} = 0$.

Ultimately for the machine to be stable, δ must stop changing as the synchronous speed is reached and at this final equilibrium point also $\frac{d\delta}{dt} = 0$.

Therefore, the criterion for stability is that $\frac{d\delta}{dt} = 0$. Hence,

$$\sqrt{\int_{\delta_0}^{\delta} \frac{2(P_L - P_m \sin \delta)}{P_j} \, d\delta} = 0$$

i.e.,
$$\int_{\delta_0}^{\delta} (P_L - P_m \sin \delta) \, d\delta = 0 \tag{2.36}$$

Let us consider a synchronous motor having the power-angle curve of Fig. 2.18. With the motor initially loaded with a load of power P_{L1}, the operating point is at A corresponding to a power angle δ_0. As the load on the shaft is suddenly increased to P_{L2}, the power angle swings to a value of δ_f at which instant the speed is again synchronous. From Eqn. (2.36) it follows that this system will be stable if

$$\int_{\delta_0}^{\delta_i} (P_{L2} - P_m \sin \delta) \, d\delta + \int_{\delta_i}^{\delta_f} (P_{L2} - P_m \sin \delta) \, d\delta = 0 \tag{2.37}$$

where δ_i is the power angle corresponding to the new load P_{L2}.

It can be seen that the term $(P_{L2} - P_m \sin \delta)$ is positive for value of power angle between δ_0 and δ_i and negative for those between δ_i and δ_f. Therefore, Eqn. (2.37) can be rewritten as

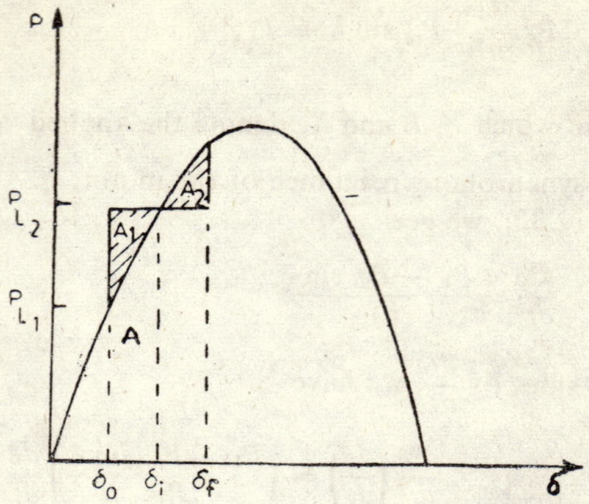

Fig. 2.18. Power-angle curve of synchronous motor and variation of load.

$$\int_{\delta_0}^{\delta_i} (P_{L_2} - P_m \sin \delta)\, d\delta = \int_{\delta_i}^{\delta_f} (P_m \sin \delta - P_{L_2})\, d\delta$$

or

$$\text{Area } A_1 = \text{Area } A_2.$$

This method of determining the transient stability of a drive system is called the equal area criterion of stability.

The equal area method gives, thus a simple indication of whether synchronism is maintained or not. Referring to Fig. 2.19, it can be seen that

(i) if area $A_2 >$ area A_1, the motor remains in synchronism and the stability is maintained,

(ii) if area $A_2 =$ area A_1, the point of operation is just stable, and

(iii) if area $A_2 <$ area A_1, the motor loses it synchronism.

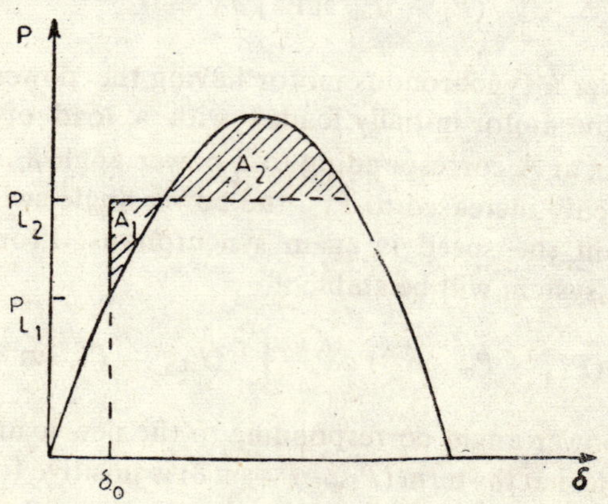

Fig. 2.19. Equal area criterion for transient stability.

It may be noted that while deriving the criterion for transient stability of the synchronous motor drive, the damping power term has been neglected. Hence, equal area criterion method gives slightly pessimistic results as regards transient stability.

Example 2-3. A synchronous motor connected to an infinite busbar is driving a load corresponding to its rated capacity, with a torque angle of 30°. If the load is suddenly increased to $\sqrt{2}$ times the rated load, determine whether or not the drive is stable.

Calculate the maximum additional load that can be thrown suddenly on the shaft of the motor without affecting the stability of the drive

$$P_{L_1} = P_m \sin \delta_1 = P_m \sin 30^\circ = P_m \times 0.5$$
$$P_{L_2} = \sqrt{2}\, P_{L_1} = \sqrt{2}\, P_m \times 0.5$$
$$= P_m \sin \delta_2$$

i.e., $$\sin \delta_2 = \frac{1}{\sqrt{2}}, \text{ so } \delta_2 = 45^\circ.$$

Referring to Fig. 2.19,

$$\text{Area } A_1 = \int_{30^\circ}^{45^\circ} \left(\sqrt{2}\, P_{L_1} - P_m \sin \delta\right) d\delta$$
$$= P_m \int_{30^\circ}^{45^\circ} (0.707 - \sin \delta) d\delta$$
$$= 0.026\, P_m$$

$$\text{Area } A_2 = \int_{45^\circ}^{(180\text{-}45^\circ)} (P_m \sin \delta - \sqrt{2}\, P_{L_1})\, d\delta$$
$$= P_m \left\{ [-\cos \delta]_{45^\circ}^{135^\circ} - 0.707\, [\delta]_{(\pi/4)}^{(3\pi/4)} \right\}$$
$$= 0.304\, P_m$$

Since area A_2 is > area A_1, the drive is stable.

Let the power angle corresponding to the safe load P_{Ls} be δ_s.

$$\text{Area } A_1 = \int_{30^\circ}^{\delta_s^0} (P_{Ls} - P_m \sin \delta)\, d\delta$$
$$= (\delta_s - 30) \times \frac{\pi}{180} P_{Ls} + P_m (\cos \delta_s - \cos 30^\circ)$$

But, we know that $P_{Ls} = P_m \sin \delta_s$.

$$\text{Therefore, Area } A_1 = P_m \left[\frac{\pi}{180} (\delta_s - 30) \sin \delta_s + (\cos \delta_s - \cos 30^\circ) \right]$$

$$\text{Area } A_2 = \int_{\delta s}^{(180-\delta_s)} (P_m \sin \delta_s - P_{Ls})\, d\delta$$
$$= P_m \left[2 \cos \delta_s - \frac{\pi}{180} (\pi - 2\delta_s) \sin \delta_s \right]$$

In order that the drive remains stable

$$\text{Area } A_1 = \text{Area } A_2$$

Equating the two expressions obtained above, we get

$$\frac{\pi}{180}(150 - \delta_s) \sin \delta_s = 0.866 + \cos \delta_s$$

Solving by trial and error, we get

$$\delta_s = 60.5°$$

Hence, maximum safe load $= P_m \sin 60.5°$

$$= 1.74\, P_{L_1}$$

So, additional load that can be thrown suddenly on the shaft = 0.74 rated load.

PROBLEMS

1. On the basis of the conventions chosen for speed and load torque in the text, show that the speed torque curves of passive torque loads are confined to the first and third quadrants only while those of active torque loads can range over all four quadrants.
2. A lift usually has some friction torques (apart from the unbalanced load torque caused by the difference in weights of the car and counterweight) which may be considered to be independent of speed.
   (a) Sketch the speed-torque curve of a fully loaded lift having significant friction torque component.
   (b) Sketch the speed torque curve of the same lift having a counterweight equal to the weight of the car.
   (c) State the conditions under which the car will remain stationary although the brakes are released and the motor not switched on ?
3. Figure 2.20 shows a weight of 1000 kg being lifted up at a velocity of 1 m/sec by means of a motor running at 960 rpm and a winch having a diameter of 0.30 m. The inertia of the motor and the winch drum are 1.6 kg-m² and 3.2 kg-m² respectively. Calculate the total load torque of the system referred to the motor shaft.
4. A horizontal conveyer belt moving at a uniform velocity of 1 m/sec. transports load at the rate of 50,000 kg/hour. The belt is 180 m long and is driven by a 960 rpm motor.

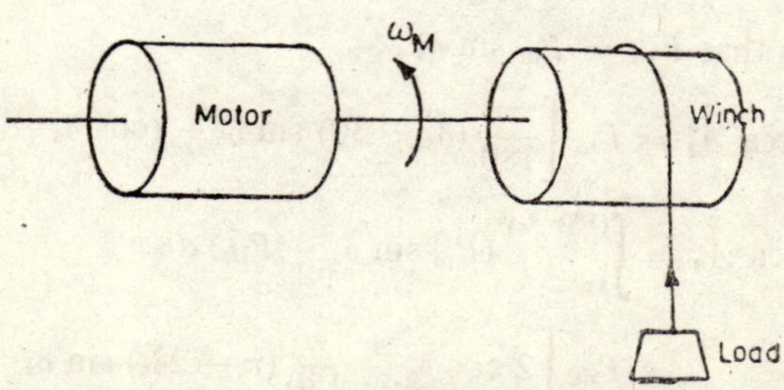

Fig. 2.20. Drive system for Problem-2.

(a) Determine the equivalent rotational inertia at the motor shaft.

(b) Calculate the required braking torque at the motor shaft to stop the belt at a uniform rate in 10 secs.?

5. Determine whether the points of equilibrium P_1 and P_2 in Fig. 2.21(a) and P_3 and P_4 in Fig. 2.21(b) are stable or not. Explain, with reasons.

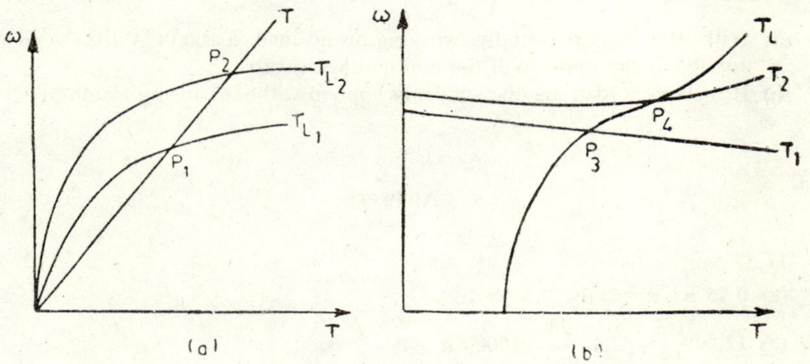

Fig. 2.21. Motor and load torque curves for Problem-4.

6. The complete speed-torque characteristic of a squirrel cage induction motor is to be obtained experimentally. Explain whether it is possible to get it by using the following loads:

   (a) Pulley and spring balances arrangement for which the torque may be assumed independent of speed.

   (b) Separately excited dc generator supplying fixed resistance load.

   (c) dc shunt generator feeding a fixed resistance load.

   (d) Fan type of load for which the torque may be assumed to vary proportionally to the square of the speed.

7. Sketch the seven possible cases of the load torque curve and motor torque curve (as shown in Fig. 2.17) with torque in the x-axis and speed in the y-axis, and determine the stability of each.

8. A 200 kW, 2300 V, 3 phase, 50 Hz, 28 pole synchronous motor is directly connected to a large power system and is driving a load. The motor has the following characteristics:

   Inertia of the motor plus load = 450 kg-m^2

   Synchronizing power = 10.0 kW/elec. deg.

   Damping torque = 2400 N-m/mech. rad./sec.

   (a) Write down the equation representing the electrodynamic oscillation of the machine.

   (b) Determine (i) the undamped natural frequency and the natural period of oscillations, and (ii) the damped natural frequency of oscillations.

   (c) Rated load is suddenly thrown on the motor shaft at a time when it is operating on steady state; but on no load. Investigate the nature of ensuing oscillations.

9. A synchronous motor is driving a reciprocating compressor, which requires a torque that varies periodically about a steady average value.

(a) Using a linear analysis, write down the electromechanical equation of the drive, when the harmonic torque component of the variable shaft torque is represented as $T_{Lm} \sin \omega_L t$.

(b) Determine the maximum amplitude of oscillation in phasor form.

10. A 1000 kW, 3 phase, 50 Hz, 600 rpm synchronous motor has a power angle curve under transient conditions represented by 3.00 sin δ, where the amplitude is in percent on a 1000 kW base.

(a) With the motor initially working on no load, a 1000 kW load is suddenly applied at the shaft. Will the motor remain stable ?

(b) How large a load may be suddenly applied without causing instability ?

Answers

3. 11.57 Nm,
4. (a) 0.25 Kg-m^2; (b) 2.5 Nm
8. (a) $12.59\dfrac{d^2\delta}{dt^2}+67.14\dfrac{d\delta}{dt}+10000\,\delta=0$

   (b) (i) $f_n = 4.485$ Hz; $T_n = 0.222$ sec

   (ii) $f_d = 4.464$ Hz.
10. (a) Yes (b) 2.15 p.u.

Chapter 3

Characteristics of dc Motors

3.1 Basic Relations

The following three fundamental equations pertaining to dc machines are quite well known:

$$V = E \pm I_a R_a, \tag{3.1}$$

$$E = \frac{1}{2\pi} \cdot \frac{P}{A} \phi Z \omega \text{ volts}, \tag{3.2}$$

and

$$T = \frac{1}{2\pi} \cdot \frac{P}{A} \phi Z I_a \text{ newton-metre}, \tag{3.3}$$

where V—voltage at the terminals of the machine, volts

E—induced emf in the armature, volts

I_a—armature current, amps

R_a—armature resistance, ohms

P—number of poles in the machine,

A—number of parallel paths in the armature,

ϕ—flux per pole, Weber

Z—number of armature conductors,

ω—angular speed of rotation of the armature, radians/sec

T—torque developed by the armature, N-m.

The positive sign in Eqn. (3.1) is applicable for a motor, while the negative sign for a generator.

3.2 Basic Characteristics

There are three basic characteristics for any dc motor, viz., speed vs. armature current characteristic; torque vs. armature current and speed vs. torque characteristic.

3.2.1 Basic Characteristics of dc Shunt Motors

The speed vs. armature current characteristic can be predetermined from Eqns. (3.1) and (3.2)

$$\omega = \frac{V - I_a R_a}{K_e \phi}, \tag{3.4}$$

where K_e is a constant for the machine. When ϕ is constant, ω can be expressed in the form $A - BI_a$, where $A = \dfrac{V}{K_e\phi}$ and $B = \dfrac{R_a}{K_e\phi}$ are also constants for the machine under study. Hence, $\omega = f(I_a)$ should be a straight line. However, due to the armature reaction, the effect of which increases with increase in armature current, the magnitude of denominator of Eqn. (3.4) reduces with increase in I_a, but to a less extent than the numerator and, therefore, has the shape shown in Fig. 3.1.

The torque vs. armature current characteristic can be obtained from Eqn. (3.3), i.e., $T = K_t\phi_a$, where K_t is a constant for the machine. The curve with dotted line in Fig. 3.2 represents this relationship, the deviation from an ideal straight line, especially at higher values of I_a being due to the effect of armature reaction, viz., the reduction in the magnitude of flux per pole. Since $I_a = I_{load} + I_f$, the actual curve will be as shown by firm line in Fig. 3.2. This curve has an interception in the x-axis given by the magnitude of the no load current of the motor.

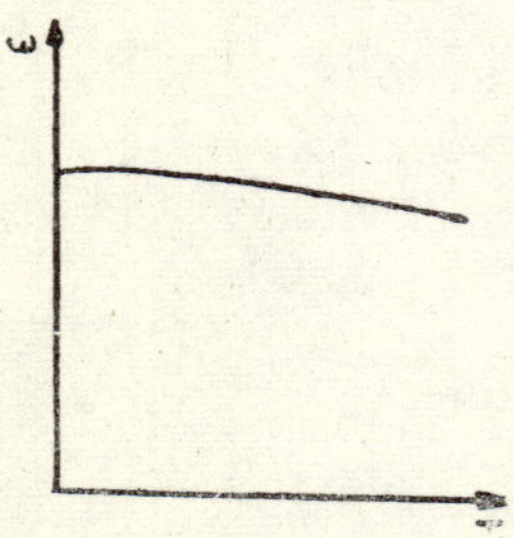

Fig. 3.1. Speed-current curve of shunt motor.

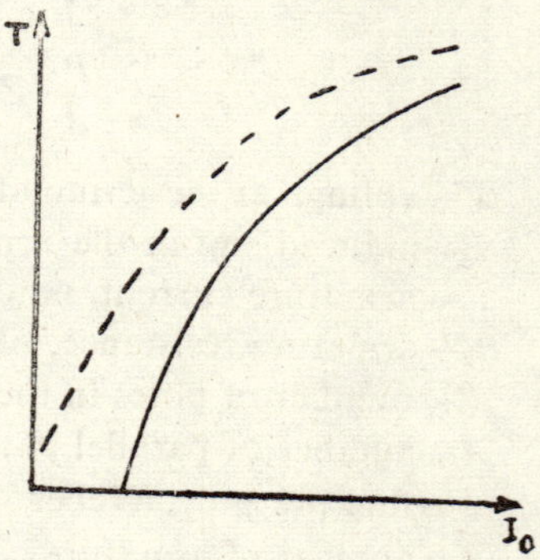

Fig. 3.2. Torque-current curve of shunt motor.

The relationship between speed and torque can be determined by using the above two characteristics. Alternatively, from Eqns. (3.3) and (3.4), we have

$$\omega = \frac{V}{K_e\phi} - \frac{R_a \cdot T}{K_e K_t \phi^2} \tag{3.5}$$

$$= A_1 - B_1 T, \tag{3.6}$$

where A_1 and B_1 are constants.

Neglecting the effect of armature reaction, $\omega = f(T)$ will be linear. Due to armature reaction, ϕ no longer remains constant for different values of I_a and hence the speed-torque characteristic attains the shape shown in Fig. 3.3.

3.2.2 *Basic Characteristics of dc Series Motors*

ω-I_a characteristic: The armature current itself acts as the exciting current for series motors

$$\omega = \frac{V - I_a R_a}{K_e \phi} \simeq \frac{V}{K_e \phi} \propto \frac{V}{I_a}$$

or $$\omega I_a \propto V \tag{3.6}$$

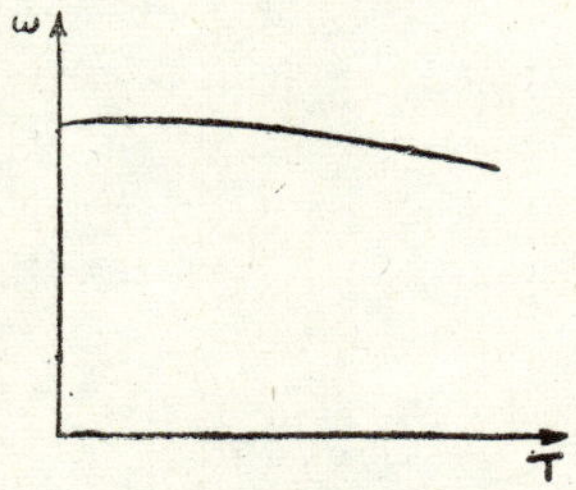

Fig. 3.3. Speed-torque curve of shunt motor.

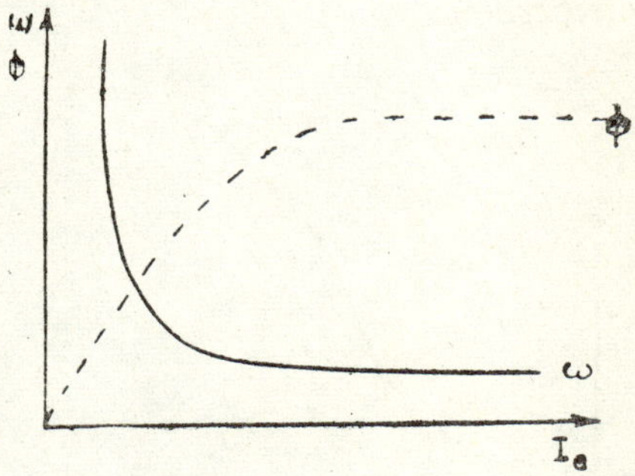

Fig. 3.4. Speed and flux per pole vs. current curve of series motor.

It follows from the above that the speed of a dc series motor is approximately inversely proportional to flux per pole or load current, as depicted in Fig. 3.4.

T-I_a characteristic: With small values of load current, the magnetic circuit remains unsaturated and

$$\phi = K_1 I_a, \text{ where } K_1 \text{ is a constant}$$

and $$T = K_t \phi I_a = K_1 K_t I_a^2 = K_2 I_a^2, \text{ where } K_2 \text{ is another constant.} \tag{3.7}$$

Hence, the initial portion of the torque-armature current characteristic is given by a parabola passing through the origin.

With larger values of load current, the magnetic circuit becomes saturated and the flux per pole more or less remains constant, irrespective of changes in load current. Therefore, at higher values of load current $T \propto I_a$. Thus, the torque-current characteristic of a series motor is of the shape in Fig. 3.5.

ω-T characteristic: The speed-torque characteristic can be derived by using the above two characteristics, by eliminating armature current.

From Eqn. (3.5), we know that

$$\omega = \frac{V}{K_e \phi} - \frac{T R_a}{K_1 K_e \phi^2}$$

Eqn. (3.7), which is valid for low values of load current can be expressed as

$$T = K_3 \phi^2 \qquad \text{or} \qquad \phi = (T/K_3)^{1/2} \tag{3.8}$$

Hence, as long as the magnetic circuit remains unsaturated

$$\omega = \frac{V}{K_e\sqrt{\frac{T}{K_3}}} - \frac{K_3\phi^2 R_a}{K_t K_e \phi^2} \tag{3.9}$$

$$= \frac{C_1}{\sqrt{T}} - D_1 \tag{3.10}$$

where C_1 and D_1 are two constants; i.e., the shape of the ω-T characteristic will be hyperbolic.

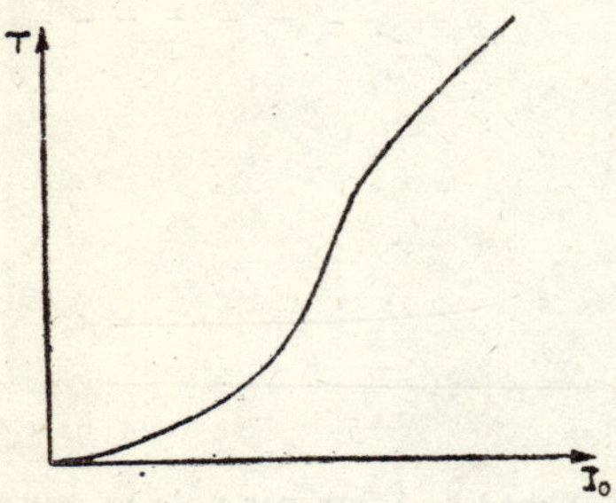

Fig. 3.5. Torque vs. current of series motor.

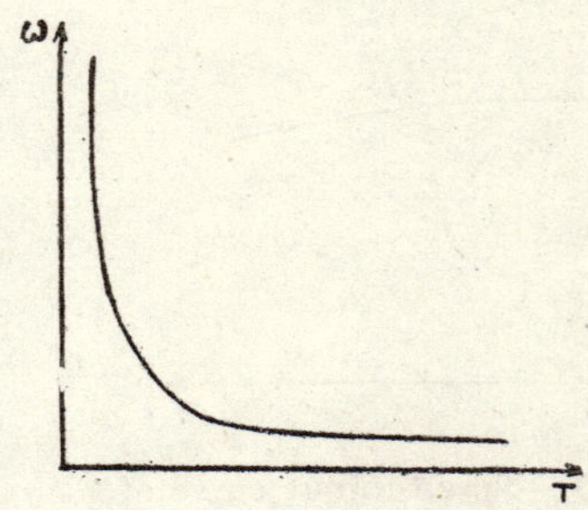

Fig. 3.6. Speed vs. torque of series motor.

For higher values of load current ϕ remains more or less constant and, therefore, speed will be given by

$$\omega = C_2 - D_2 T \tag{3.11}$$

where C_2 and D_2 are constants for the machine; i.e., the characteristic will be linear in nature. Thus, the speed-torque characteristic of the series motor takes the form shown in Fig. 3.6.

Example 3-1. A dc shunt motor is connected to constant voltage mains and drives a load torque which is independent of speed. Prove that, if E (induced emf) $> \frac{1}{2}V$ (supply voltage), increasing the air gap flux per pole decreases the speed of the motor, while, if $E < \frac{1}{2}V$ increasing the air gap flux per pole increases the speed.

Solution: $V = E + I_a R_a$

$$= K_e\phi\omega + I_a R_a$$

Hence $$\omega = \frac{V - I_a R_a}{K_e\phi}$$

i.e., $$\omega = \frac{V\phi - \phi I_a R_a}{K_e\phi^2}$$

$$= \frac{V\phi - (T/K_e)R_a}{K_e\phi^2} \qquad (\because\ T = K_e\phi I_a)$$

Both V and T are given as constants. R_a is also a constant.

Hence, $$\frac{d\omega}{d\phi} = -\frac{V}{K_e\phi^2} + \frac{2TR_a/K_e}{K_e\phi^3}$$

If $\frac{d\omega}{d\phi} > 0$, an increase in flux per pole will cause an increase in speed

$$\frac{d\omega}{d\phi} > 0 \text{ if } \frac{2TR_a/K_e}{K_e\phi^3} > \frac{V}{K_e\phi^2}$$

$$\text{if } \frac{2TR_a}{K_e\phi} > V$$

$$\text{if } 2I_aR_a > V$$

$$\text{if } 2V - 2E > V$$

$$\text{if } V > 2E \qquad \text{or} \qquad V/2 > E$$

i.e., $$\text{if } E < V/2$$

If $d\omega/d\phi < 0$, an increase in flux per pole will cause a decrease in speed. From above, it is obvious that $d\omega/d\phi < 0$, if $E > V/2$.

3.2.3 *Effect of Impulsive Changes in Supply Voltage*

Let us assume that the change in supply voltage occurs so suddenly that during that period the motor speed cannot change because of mechanical inertia and the magnetic flux cannot vary due to electromagnetic inertia. Since armature current $I_a = \frac{V - E}{R_a}$, the variation in I_a due to change in V depends only on E.

(i) *Separately excited motor*: With a change in applied voltage, the field flux would not change since the field is excited from a separate source.

Let the armature current change from

$$I_{a_1} = \frac{V_1 - E}{R_a} \text{ to } I_{a_2} = \frac{V_2 - E}{R_a}$$

Relative variation in armature current

$$\frac{I_{a_2} - I_{a_1}}{I_{a_1}} = \frac{V_2 - V_1}{V_1 - E}$$

or

$$\frac{\Delta I_a}{I_{a_1}} = \frac{\Delta V}{V_1 - E} \tag{3.12}$$

Assuming an armature resistance drop of 5 per cent,

$$V_1 - E = I_{a_1}R_a = \frac{5}{100}V_1,$$

and, hence

$$\frac{\Delta I_a}{I_{a_1}} = 20\left(\frac{\Delta V}{V_1}\right) \tag{3.13}$$

Thus, a sudden change of 1 per cent in the magnitude of supply voltage causes a change of 20 per cent in armature current. If, for example, the voltage increases by 5 per cent, the armature current increases by 100 per cent and will worsen commutation.

(ii) *Shunt motor*: The inductance of the field winding opposes any instantaneous change in exciting current that is likely to take place due to a change in applied voltage. Therefore, the magnitude of induced emf E initially remains unchanged. Due to, say, an increase in the difference between applied voltage V and induced emf, the current in the armature increases. But, this sudden increase dies down within a short time because of the increase in magnitude of E due to an increase in ϕ caused by the increased value of the field current.

(iii) *Series motor*: If applied voltage V were to increase, the armature current tends to increase until E retains its initial value. But, since an increase in armature current is associated with an increase in flux, the induced emf also increases which finally brings the current to its initial value. Thus, series motors are less sensitive to sudden changes in supply voltage.

3.2.4 *Effect of Fluctuation in Load Torque*

In the case of both separately excited and shunt motors, a change in load torque demands a proportional change in the magnitude of the armature current.

In order to determine the effect of change in load torque on the armature current of a series motor, let us consider the basic torque equation once again.

$$T = K_t \phi I_a \tag{3.14}$$

Differentiating the above, we have

$$dT = K_t (\phi dI_a + I_a d\phi) \tag{3.15}$$

Hence,

$$\frac{dT}{T} = \frac{dI_a}{I_a} + \frac{d\phi}{\phi} \tag{3.16}$$

or for small deviations in torque, current and flux

$$\frac{\Delta T}{T} = \frac{\Delta I_a}{I_a} + \frac{\Delta \phi}{\phi} \tag{3.17}$$

From the above expression it is clear that for the same percentage increase in electromagnetic torque (as required by the increased load torque) the percentage increase in armature current will be less for series motor than shunt motors, because of the simultaneous increase in magnetic flux in the case of series motors. From the point of view of reduced fluctuations in supply voltage, this characteristic of the series motor is an advantage.

3.2.5 *Basic Characteristics of Compound Motors*

Cumulatively compounded motor: Obviously, the characteristics of such a motor lies somewhere in between those of shunt and series motors. It is to be noted that the effect of series winding is negligibly small at low values of armature current. All the three commonly used characteristics are shown in Fig. 3.7.

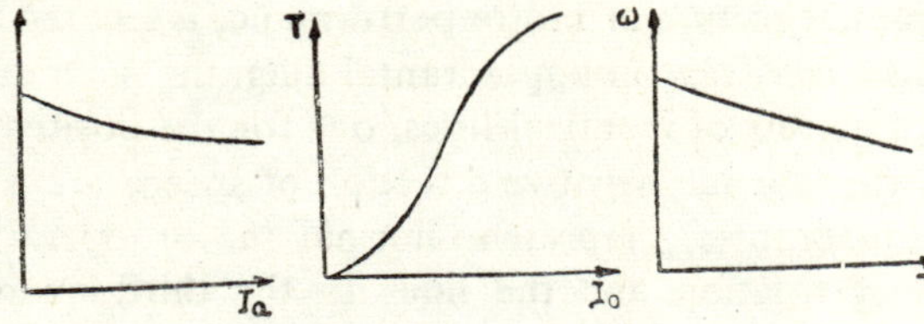

Fig. 3.7. Characteristics of compound motor.

Differentially compounded motor: Since the series field winding is connected in such a way as to oppose the effect of shunt field winding, it is possible to obtain even a rising speed characteristic with increase in load. However, this type of motor is rarely used in industry, since it doesn't ensure a stable operation of the drive with constant torque loads.

Another disadvantage of the motor is its performance during either starting or overload. Under these conditions of heavy armature currents, the series field flux can exceed the main shunt field flux, causing the polarity of the motor to change. Since the motor is directly connected with the supply, a short circuit occurs.

3.3 Modified Speed Torque Characteristics of DC Shunt Motors

The most simple means of obtaining a variety of speed torque characteristics of dc shunt motors is to introduce additional resistances either in the armature circuit or field circuit.

3.3.1 Introduction of Armature Series Resistances

Figure 3.8 shows the circuit diagram. The speed is given by

$$\omega = \frac{V}{K_e\phi} - \frac{(R_a + R)T}{K_eK_t\phi^2} \tag{3.18}$$

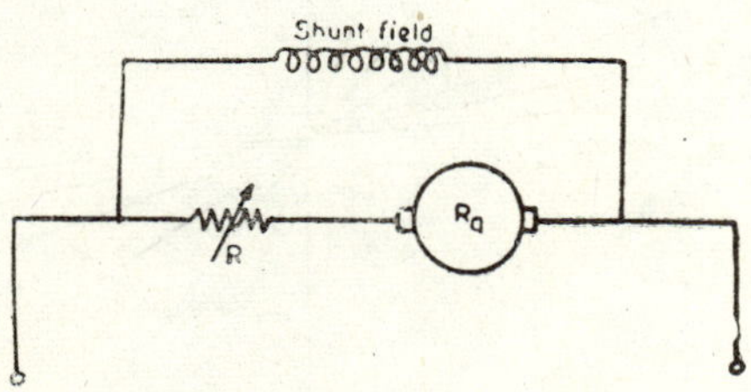

Fig. 3.8. Armature series resistance for shunt motor.

Neglecting the effect of saturation, the above equation represents the speed-torque characteristic and is depicted by a straight line (whose slope is a function of $(R + R_a)$ intersecting the ω axis at a value given by $V/K_e\phi$. For different values of additional resistance R, the speed-torque curves will be a family of straight lines.

A complete picture of motor performance is secured by plotting the speed torque characteristics on a quadrantal diagram, such as in Fig. 3.9. In this diagram, two sets of identical lines, one for the positive (or forward) speed and the other for the negative direction of speed, are shown. The lines in the first quadrant (I) represent normal motor operation in the positive direction of rotation and the lines in the third quadrant (III) represent normal motor action, but in that opposite direction. Quadrants II and IV represent braking action, in the *motor torque is opposite in sign to that of the speed.* Those portions of the upper set of lines which extend into the second quadrant depict generator action, or simply regeneration. Those portions of the same set of lines which extend into the fourth quadrant represent plugging. Plugging (reverse current braking) is a type of braking action caused by reversing the armature polarity so that the motor torque acts in a sense opposite to that of speed, to cause a reduction in its value.

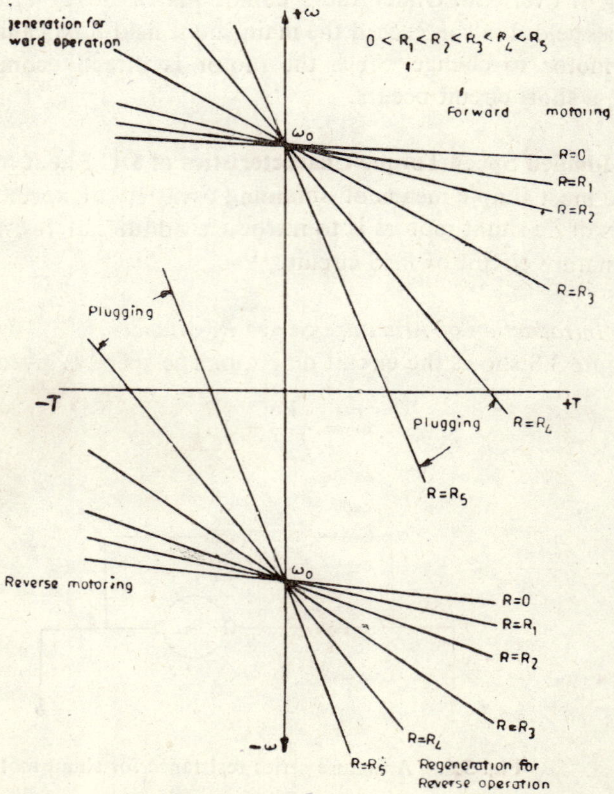

Fig. 3.9. Speed-torque curves for different values of armature series resistance.

3.3.2 Variation of Field Current

The circuit diagram is as shown in Fig. 3.10.

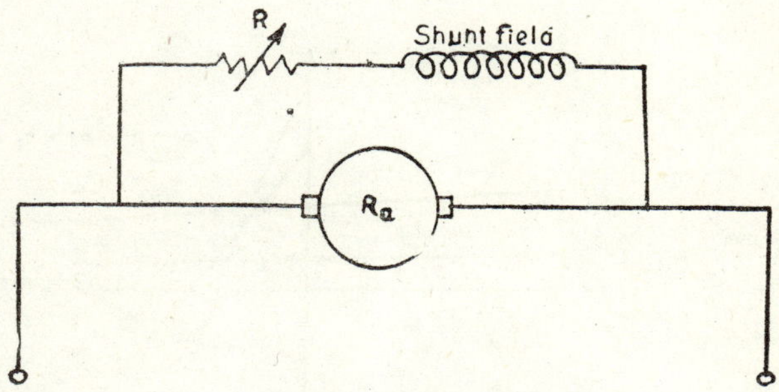

Fig. 3.10. Field resistance variation in shunt motor.

From the equation

$$\omega = \frac{V}{K_e\phi} - \frac{R_a T}{K_e K_t \phi^2},$$

it is clear that for the same torque on the motor, various speeds of operation are possible, by varying field flux. Variation in field flux can be achieved by introducing a variable resistance in the field circuit.

The motor speed torque curves for various values of field resistance will appear as in Fig. 3.11, if the effect of armature reaction is neglected. It may be seen that both the no load speed (represented by the intercept on the speed axis) and the slope of the speed torque curve change with the variation in the field circuit resistance. If the magnitude of resistance added in series with the field is considerably large, the field flux gets reduced significantly and hence the no load speed becomes quite high and the slope of the speed torque curve also increases as shown by the two lines for $R = R_3$ and R_4 in Fig. 3.11. However, in actual practice, due to the effect of armature reaction, which becomes dominant when field flux is weakened considerably, the operation of the motor may become unstable and commutation of the motor very bad. In fact, the maximum permissible value of the armature current will set a limit to the magnitude of the change in field current and hence in the additional resistance in the field circuit.

Quadrants I and III again represent normal motor operation in the positive direction of rotation and in the negative direction respectively. Quadrant II indicates the condition where the load drives the motor in either forward or backward direction faster than the ideal no load speed. In other words, quadrant II (both for the upper and lower set of lines) represents regenerative braking regime of the motor.

3.4 Modified Speed Torque Characteristics of dc Series Motors

A greater variety of speed-torque characteristics is obtainable with a series motor than with a shunt motor, by means of different connections for the field and armature circuits together with suitable series and shunt resistors. The commonly used connections are (i) series resistance, (ii) shunted motor connection, and (iii) shunted armature connection.

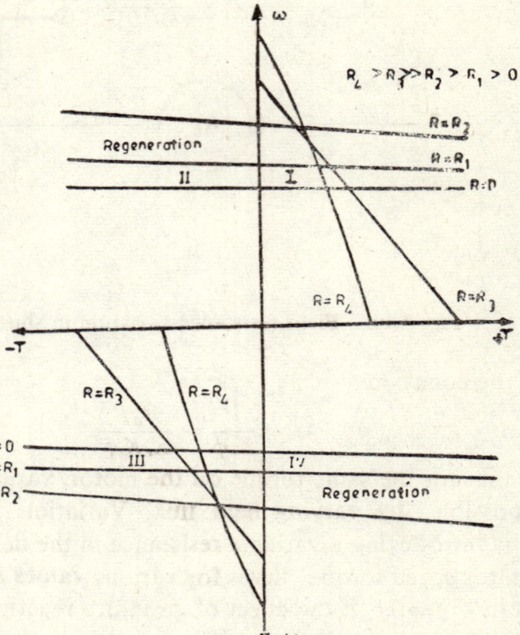

Fig. 3.11. Speed-torque curves for different field circuit resistances.

3.4.1 Series Resistance

The obvious method of varying the speed of a series motor is to use a series resistance. The schematic diagram is shown in Fig. 3.12 and the resulting speed-torque curves for several values of resistances in Fig 3.13. It may be noted that the no load motor speed corresponding to small no load torques may be excessively high even with a fairly large series resistance. With an increase in the value of the series resistance, the torque at any given speed is reduced. A considerable increase in the series resistance can bring the curve into the fourth quadrant, since the induced emf can become

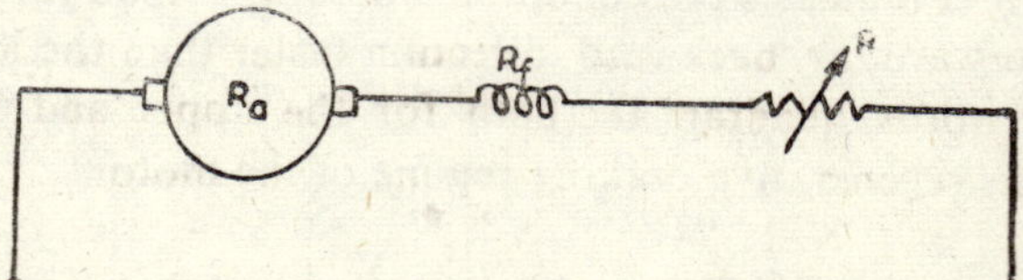

Fig. 3.12. Variation of motor circuit resistance in a series motor.

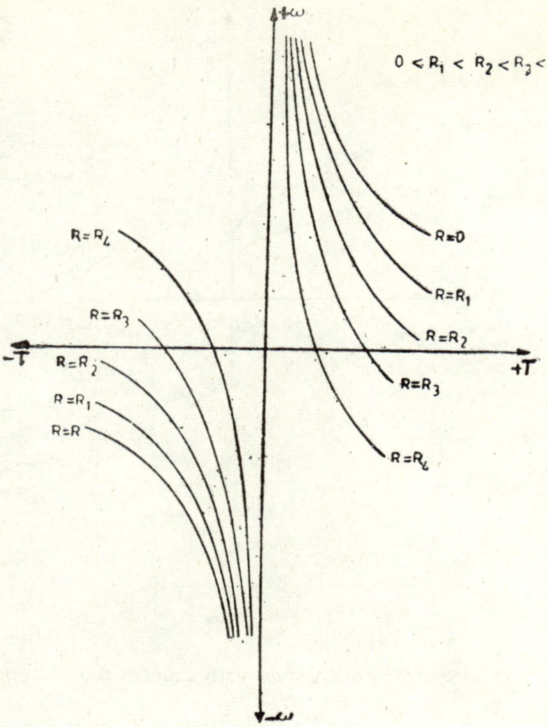

Fig. 3.13. Speed torque curve for various circuit resistances of series motor.

negative with a large drop in the motor circuit resistance and hence causing a change in direction of rotation.

Also, the curves for operation in the reverse direction (third quadrant) are identical with those in the first.

3.4.2 Shunted Motor Connection

The circuit shown in Fig. 3.14, in which resistors are connected both in series and parallel with a series motor, is called the shunted motor connec-

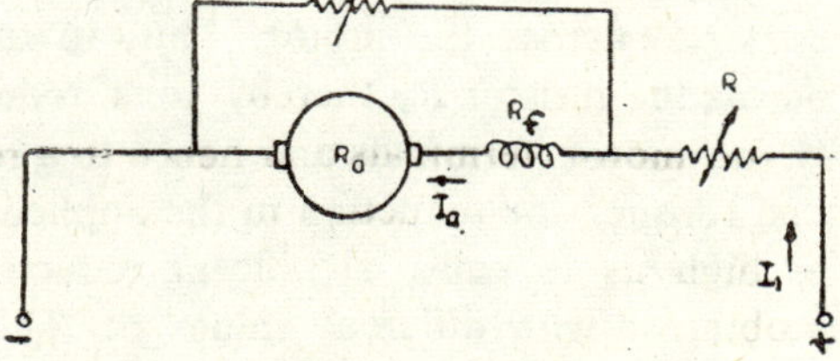

Fig. 3.14. Shunted motor connection of series motor.

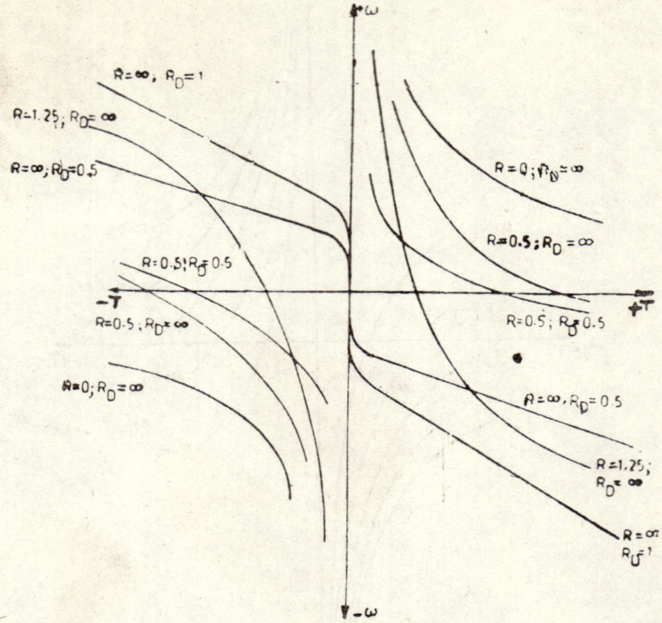

Fig. 3.15. Speed-torque curves with shunted motor connection.

tion. The basic curve is marked $R_D = \infty$; $R = 0$ in Fig. 3.15. With an increase in the value of series resistance ($R_D = \infty$, $R = 0.5$ p.u.) the torque at any given speed is reduced. A considerable reduction in the starting torque is also observed. A further addition in the series resistance ($R_D = \infty$, $R = 1.25$ p.u.) makes the curve enter deep into the fourth quadrant. It must be noted that these cuvres must be exactly similar in nature to those obtained in the section 3.4.1, since with $R_D = \infty$, the circuit shown in Fig. 3.14 will be exactly the same as in Fig. 3.12. If resistances are connected both in series and parallel with the series motor ($R_D = 0.5$ p.u., $R = 0.5$ p.u.), for a given torque, a reduction in speed, compared with the curves for the motor connected with series resistances alone ($R_D = \infty$, $R = 0.5$ p.u.), can be observed. This is due to the fact that for the same value of torque, i.e., for the same value of armature current I_a, the current drawn from the supply I_1 has necessarily to be larger in magnitude because of the presence of a shunt resistance across the motor. This, in turn, leads to a larger voltage drop across the resistor R, thereby to a reduction in the voltage applied across the motor terminals and hence to a reduction in speed. At larger values of torque, the reduction in the applied voltage to the motor may not be as high as to cause significant reduction in speed. All the characteristics obtained with different values of R_D and R tend to have infinitely large speeds at small values of torque and, therefore, do not enter into the second quadrant. Once again, it may be seen that the corresponding curves for operation of the motor in the reverse direction, shown in the third quadrant, are identical with those in the first. A careful study of the above two conditions will indicate that a series motor will not be able to build up emf as a generator, if the armature and field circuit is simply shunted across an external resistor. The connections of the field with respect to the armature have to be reversed. The two characteristics corresponding to $R = \infty$, shown in Fig. 3.15, have necessarily been obtained after connecting the field in a reverse manner to that used for the other curves in the same figure. The salient features of the above characteristics are:

(i) Being a connection for generator operation, it is not possible to obtain a motoring torque.
(ii) Even, braking torque is developed only at speeds above some minimum value. This is due to the fact that the torque can appear only when current flows in the machine, which in turn requires a minimum emf to be induced and hence a minimum speed of rotation of the machine. The magnitude of the minimum speed becomes larger as the resistance of the machine circuit increases, since more emf has to be induced to overcome the voltage drop in the machine circuit and to pass a current.
(iii) The speed at a given value of braking torque increases with an increase in the resistance of the machine circuit, since the voltage drop in the circuit increases necessitating a larger induced emf to sustain the flow of current.

The two remaining curves in Fig. 3.15 correspond to the situation when the motor is disconnected from the line, i.e., $R = \infty$ and when the machine is functioning as a self-excited series generator through the shunt resistor R_D. The motor remaining disconnected from the supply, no emf and, hence, no torque can be developed at standstill conditions of the machine and in order to develop torque the machine must get itself induced with an emf as a self-excited dc generator. The following two conditions necessary for the self-excitation of a dc machine have to be fulfilled:

(i) The total armature and field circuit resistance must be less than the critical value.
(ii) The field must be connected in such a way that the field current will aid the residual magnetism and the emf build up process takes place.

3.4.3 Shunted Armature Connection

In this connection a resistance is placed in parallel with the armature only. As the value of the divertor resistance decreases, the curves move downward as well as to the left into the second quadrant. The braking torque produced by this connection is due to the fact that the armature can now regenerate through the divertor resistance, simultaneously drawing excitation current from the line.

The connection that does effectively limit the no load speed and reduce it to any desired value is the shunted armature connection shown in Fig. 3.16. The field current does not tend to become zero even as the armature current tends to zero value and has a minimum value for quadrant I operation determined by the sum of the resistors R_f, R and R_D.

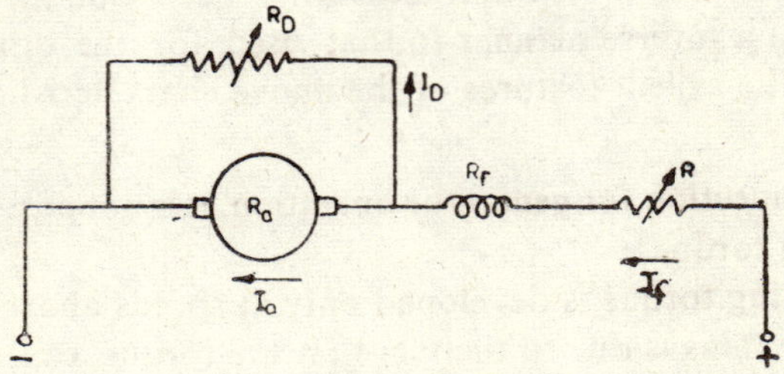

Fig. 3.16. Shunted armature connection.

(Note that $I_f = I_a + I_D = I_a + \dfrac{V - I_f \overline{R_f + R}}{R_D} = I_a + \dfrac{V}{R_D} - \dfrac{I_f \overline{R_f + R}}{R_D}$

$$\because \quad I_f = \left(I_a + \frac{V}{R_D}\right) \Big/ \left(1 + \frac{\overline{R + R_f}}{R_D}\right) \tag{3.19}$$

Therefore, the no load speed is limited to a desired quantity. This speed at which the motor torque becomes zero, obviously occurs when the armature induced emf equals the drop through the divertor resistance, i.e., when $(V - I_f \overline{R_f + R}) = E$. When the load drives the armature of the motor at a speed greater than its no load speed, E becomes greater than $(V - I_f \overline{R_f + R})$, the armature current reverses and its effect is to increase the voltage across R_D. This results in a decrease in field current and at some point such as the motor torque reaches a maximum value and then decreases as the speed becomes greater. If the load torque exceeds this maximum torque developed by the motor the speed may then increase to excessive values.

The characteristics corresponding to different values of R_D and R are depicted in Fig. 3.17. The corresponding curves for operation in the reverse direction are also shown in the same figure.

Example 3-2. Explain how would you design (choose the resistances required) a shunted armature circuit for a given motor so that the speed torque characteristic passes through (i) the speed ω_0 at zero torque and (ii) The torque T_0 at zero speed. Assume that these conditions are within the capabilities of the motor electromagnetically.

Solution: The shunted armature circuit is as shown in Fig. 3.16.

(i) Since the torque is zero, $I_a = 0$ and hence

$$I_D = I_f \text{ and}$$

$$V = I_f (R + R_f + R_D); \tag{3.20}$$

$$E = V - I_f (R + R_f) = I_f R_D \tag{3.21}$$

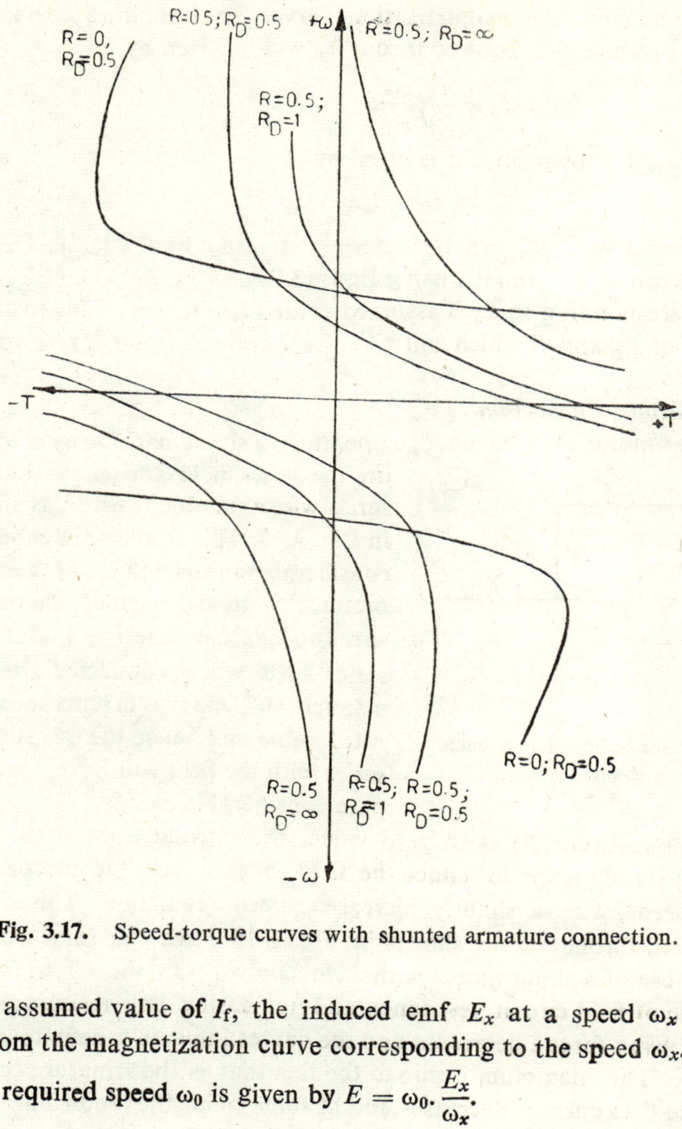

Fig. 3.17. Speed-torque curves with shunted armature connection.

For any assumed value of I_f, the induced emf E_x at a speed ω_x can be found from the magnetization curve corresponding to the speed ω_x. Hence, E at the required speed ω_0 is given by $E = \omega_0 . \dfrac{E_x}{\omega_x}$.

From Eqn. (3.21), $R_D = \dfrac{E}{I_f}$.

Knowing R_D, using Eqn. (3.20) R can be determined. Thus, depending on the value of I_f chosen, it is possible to find the values of R_D and R which will give the required speed at zero torque.

(ii) When the speed is zero, $E = 0$ and hence

$$I_a R_a = I_D R_D \tag{3.22}$$

For any assumed value of I_f, the corresponding value of E_x at a speed ω_x can be found from the magnetization curve. The armature current I_a necessary to produce the required torque T_0 will be given by

$$I_a = \frac{T_0 . \omega_x}{E_x}.$$

Knowing I_a, I_D is known since it is given by

$$I_D = I_f - I_a.$$

Once I_D is found out, R_D can be determined using Eqn. (3.22). Having found R_D, R can be determined using Eqn. (3.20).

Thus, corresponding to each assumed value of I_f, it is possible to determine values of R_D and R which will give the required torque at zero speed.

3.4.4 Shunt Motor Connection

The series motor may be made to operate as a shunt machine by connecting the series field across the line, in series with a suitable resistor, as shown in Fig. 3.18. This connection enhances considerably the usefulness of the series motor, by giving various shapes of speed-torque characteristics. But the series field when connected in this manner will carry currents near to rated value and hence the resistors in series with the field will have to dissipate considerable energy.

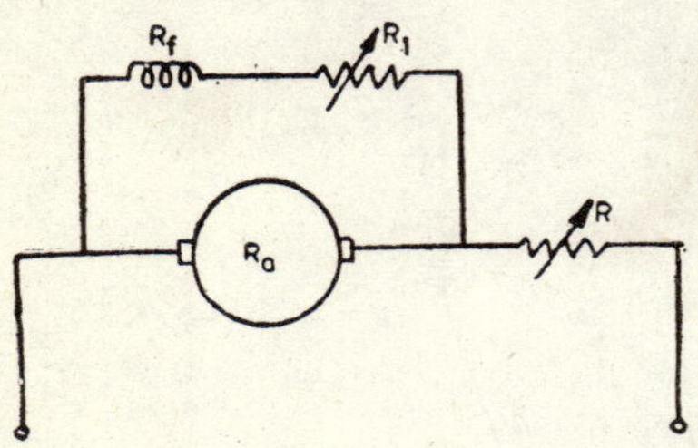

Fig. 3.18. Series motor connected as shunt.

If the line resistor (R) is of zero value, the introduction of the field resistor (R_1) simply helps to reduce the field current and the motor runs at higher speeds, with a slightly increased speed regulation. The corresponding speed torque curves shown in Fig. 3.19 have the same general nature as those of a shunt motor with field control. On the contrary, for a fixed value of field circuit resistance and increasing line resistance, the speed for a given torque decreases and the curves indicate a definite maximum torque. This maximum is due to the fact that as the armature current increases the field current decreases and at some point the product of armature current and flux per pole reaches a maximum value. If the line resistor is opened ($R = \infty$), the speed torque curve similar to that of a series generator curve, shown in Fig. 3.15, is obtained. The quadrantal diagram also indicates the corresponding curves for reverse operation of the motor.

The circuit shown in Fig. 3.20 is similar to that of Fig. 3.18 except for the addition of an armature series resistance (R_2), which is kept constant at 0.5 p.u., value for all the curves shown in Fig. 3.21. These characteristics are seen to be somewhat similar to those of Fig. 3.19, but the series armature resistance introduced increases the speed regulation and shifts the maximum torque points in to the fourth (and second) quadrants.

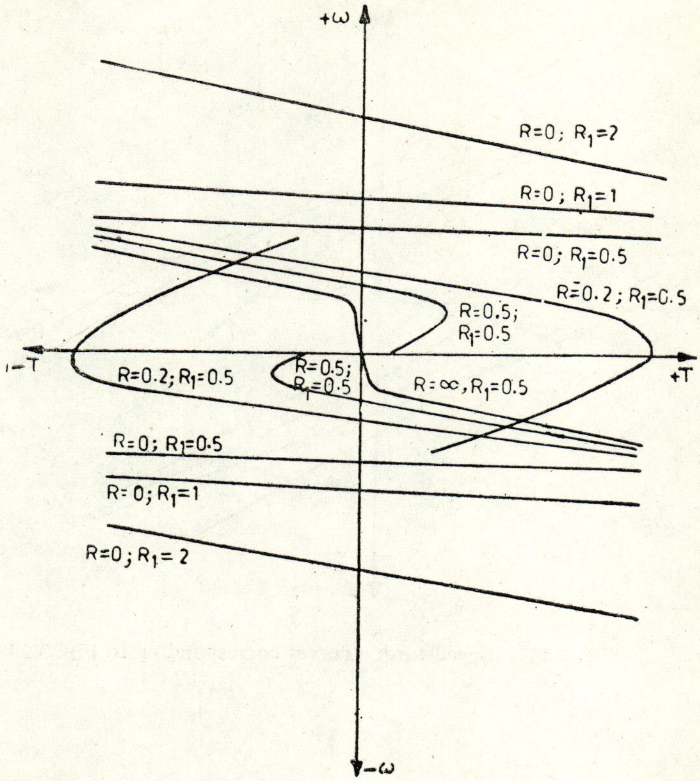

Fig. 3.19. Speed-torque curves for a series motor connected as shunt.

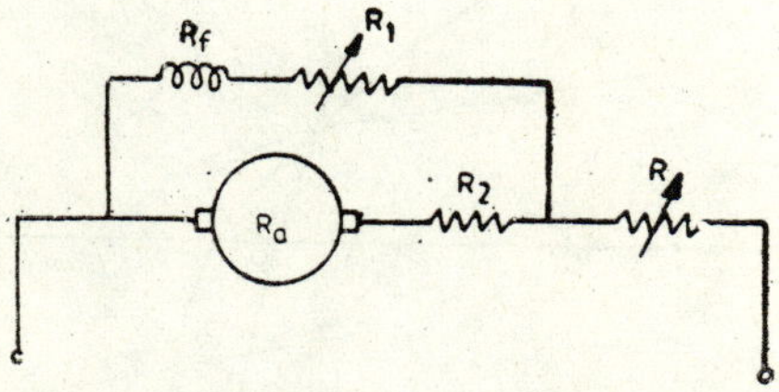

Fig. 3.20. Another variant of shunt connection of series motor.

A few more modifications of the speed-torque characteristics obtained with the series field shunted across the armature are shown in Figs. 3.22 and 3.23. In Fig. 3.22 the armature series resistance is varied, the other resistances R and R_1 being kept constant at a value of 0.5 p.u. All these curves are seen to meet at one point (corresponding to the speed at zero torque), which is the condition corresponding to zero armature current. The magnitude of this no load speed is fixed by the total resistance in

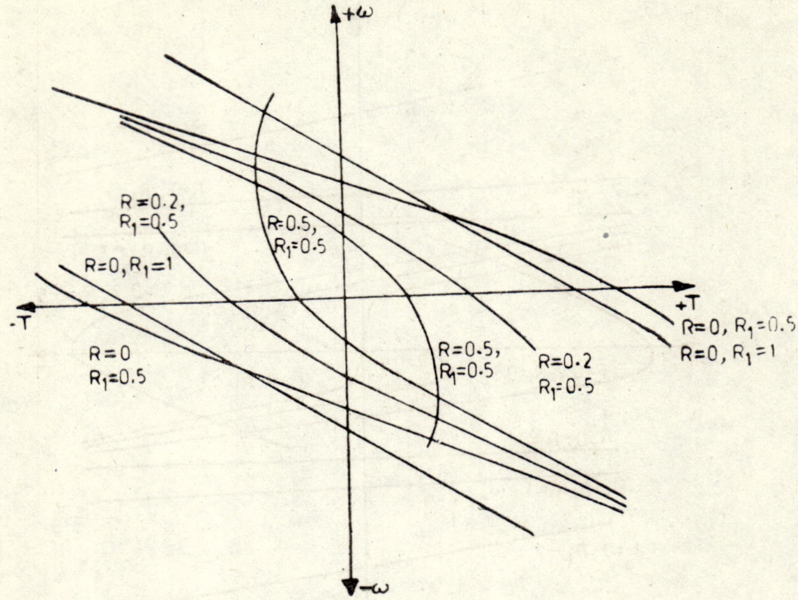

Fig. 3.21. Speed-torque curves corresponding to Fig. 3.20.

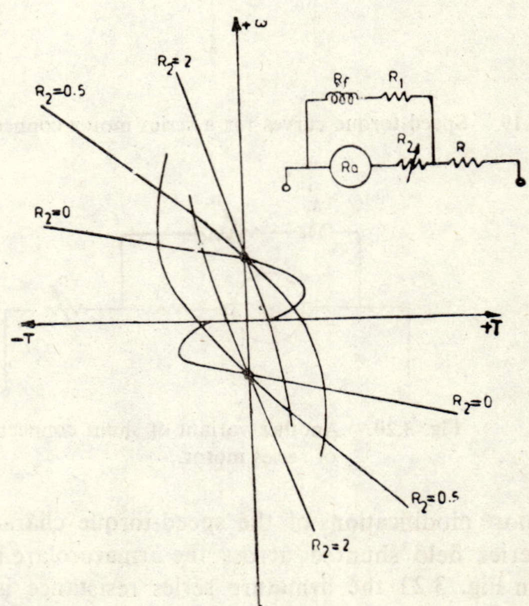

Fig. 3.22. Speed-torque curves with shunt connection armature circuit resistance alone varied.

series with the field. The starting torque is seen to increase with increase in R_2 over a small range and then decrease with further increase in R_2. The initial increase in starting torque may be explained as follows: With a small value of armature circuit resistance (say, the inherent resistance of the armature), the voltage drop in the line resistor is large and the field current relatively small. An increase in the armature circuit resistance will decrease the drop in the line resistor, thus allowing an increase in field current and hence torque. As this resistance is increased further, the decrease in armature current is more than the increase in field current causing a net reduction in the starting torque.

Figure 3.23 shows the speed torque curves obtained when the field circuit resistance is varied, keeping the other resistances R and R_2 constant at 0.5 p.u. They are similar to those of Fig. 3.22. But each curve has a different no load speed since it is determined by the field circuit resistance. The starting torque again has a maximum value for the same reasons mentioned in the earlier paragraph.

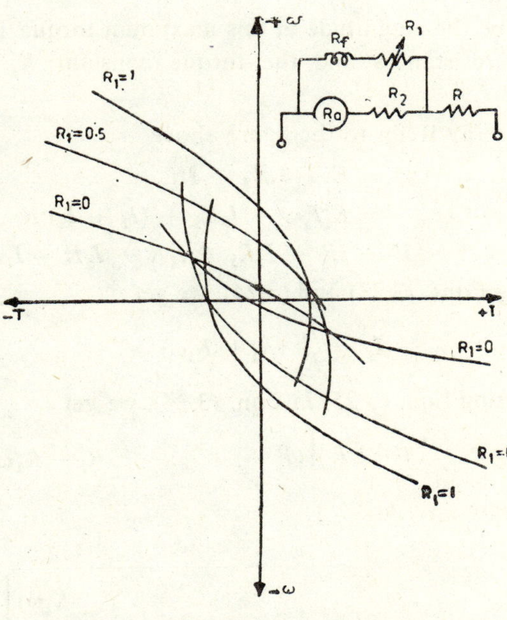

Fig. 3.23 Speed-torque curves with shunt connection-field circuit resistance alone varied.

Example 3-3. A dc series motor has been connected as shown in Fig. 3.24 for obtaining different speed-torque characteristics.

(a) Show that the motor torque passes through a maximum value when the speed of the motor is given by

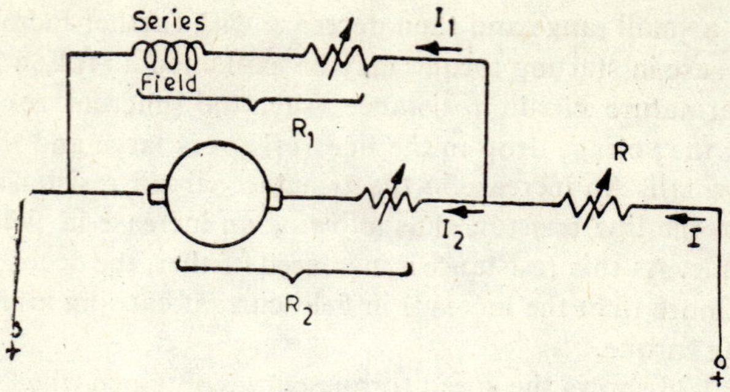

Fig. 3.24. Shunt connection of series motor for Example 3-3.

$$\omega_{T_{max}} = \frac{RR_1 - RR_2 - R_1R_2}{K_1R},$$

where K_1 is the motor constant in the voltage equation of the form $V = R_aI_a + K_1\, I_f\, \omega$.

(b) Determine the magnitude of this maximum torque in terms of applied voltage, resistances and the torque constant K_2 in the expression $T = K_2I_fI_a$.

Solution: (a) Refer to the figure above.

$$V = E + I_2R_2 + IR$$

$$= K_1I_1\omega + I_2R_2 + (I_1 + I_2)\,R \qquad (3.23)$$

Also $$V = IR + I_1R_1 = I_1R + I_2R + I_1R_1 \qquad (3.24)$$

Equating Eqns. (3.23) and (3.24), we get

$$I_2 = \frac{I_1}{R_2}\,(R_1 - K_1\omega) \qquad (3.25)$$

Substituting Eqn. (3.25) in Eqn. (3.24), we get

$V = I_1 \Big[(R +$ [illegible] $R_1 - K_1\omega)\Big],$

from where

[illegible] $K_1\omega)\Big]$ (3.26)

$T = K_2 I_1 I_2$

$$= \frac{K_2\cdot I_1^2\,(R_1 - K_1\omega)}{R_2} \qquad (3.27)$$

Substituting Eqn. (3.26) in Eqn. (3.27), we obtain

$$T = \frac{K_2\cdot(R_1 - K_1\omega)\cdot V^2}{R_2\left[(R + R_1) + \dfrac{R}{R_2}\,(R_1 - K_1\omega)\right]^2} \qquad (3.28)$$

Differentiating the above expression w.r.t. ω and equating it to zero for a maximum, we get

$$\frac{K_2V^2}{R_2}\left[\left\{(R + R_1) + \frac{R}{R_2}(R_1 - K_1\omega)\right\}^2(-K_1)\right.$$
$$\left. - 2\left\{(R + R_1) + \frac{R}{R_2}(R_1 - K_1\omega)\right\}\frac{R}{R_2}(-K_1)(R_1 - K_1\omega)\right]$$
$$= 0$$

i.e.,
$$R + R_1 + \frac{R}{R_2}(R_1 - K_1\omega) = 2(R_1 - K_1\omega)\frac{R}{R_2}$$

.e.,
$$R + R_1 = \frac{R}{R_2}(R_1 - K_1\omega)$$

i.e.,
$$RR_2 + R_1R_2 - RR_1 = -RK_1\omega$$

i.e.,
$$\omega = \frac{RR_1 - RR_2 - R_1R_2}{K_1R} \tag{3.29}$$

This is the speed at which maximum torque occurs.

(b) The maximum torque can be obtained by substituting Eqn. (3.29) in Eqn. (3.28).

$$T_{\max} = \frac{K_2 \cdot V^2\left[R_1 - R_1 + R_2 + \frac{R_1R_2}{R}\right]}{R_2\left[(R + R_1) + \frac{R}{R_2}\left(R_2 + \frac{R_1R_2}{R}\right)\right]^2}$$
$$= \frac{K_2V^2\left(1 + \frac{R_1}{R}\right)}{4(R + R_1)^2}$$
$$= \frac{K_2V^2}{4R(R + R_1)}$$

3.5 Application of Modified Characteristics

In order to illustrate the use of various possibilities of obtaining modified speed-torque characteristics, a composite set of curves for a hoisting application involving lifting and lowering heavy loads will be considered. The requirements of the load are that the motor and control must be capable of raising and lowering loads varying from no load to maximum load at low, medium and high speeds. A set of speed torque curves that satisfy the above requirements are given in Fig. 3.25. Five hoisting and five lowering positions are assumed for the hoist controller. Hoisting direction of rotation is assumed to be quadrant I operation. To obtain low and medium no load hoisting speeds, as indicated by the curves H_1 and H_2 of Fig. 3.25, the armature shunt connection (as shown in Fig. 3.16) is used. The curves H_3, H_4 and H_5 are obtained from the schematic circuit shown in Fig. 3.12 with series resistance values of R chosen to give approximately evenly spaced curves. H_5 is inherent series motor speed-torque curve with R equal to zero. From the curves shown in quadrant I, it is seen that the hoisting requirements are satisfied.

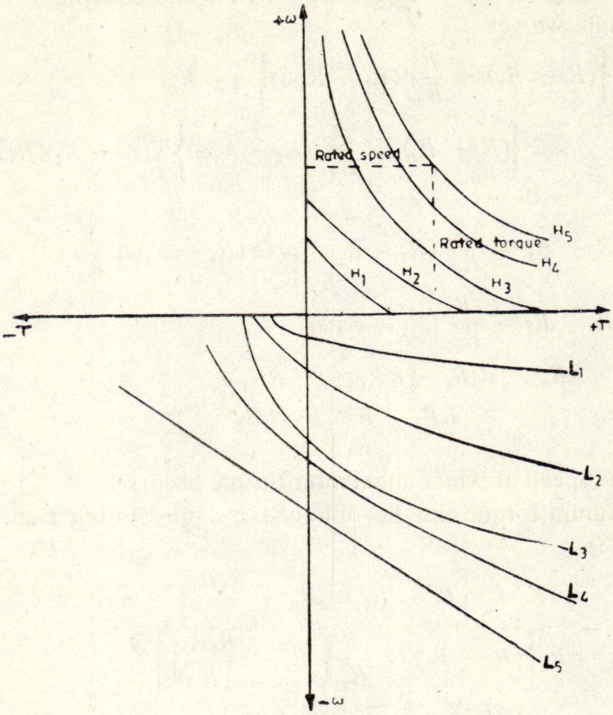

Fig. 3.25. Typical speed-torque curves desired from a series motor for hoisting applications.

To meet the requirements during lowering of loads, the connection shown in Fig. 3.20 is used and by varying R, R_1 and R_2 the different curves L_1—L_5 of Fig. 3.25 can be obtained. The high speed lowering position L_5 is secured with R equal to zero and R_2 of such a value that the curve is almost parallel to the curve L_4. An inspection of these curves shows that with no load, the empty cage can be driven down at low, medium and high speeds, as shown in quadrant III. For the overhauling conditions in the lowering direction, corresponding to quadrant IV operation, heavy loads can be lowered at low, medium and high speeds, with curves well spaced and the slopes of the curves in each position small enough so that the speed on a given lowering position does not change very much from light to heavy loads.

3.6 Direct Control of Armature-Terminal Voltage

Instead of modifying the speed-torque characteristics of dc motors in a variety of fashion, as described earlier, essentially, by having an indirect control over the applied voltage to the armature terminals (in addition to other factors), with the help of resistances inserted in high power circuits, direct control over armature-terminal voltage enables us to get suitable characteristics for various applications.

The speeds corresponding to two different armature voltages V_1 and V_2 of a dc motor are given by,

$$\omega_1 = \frac{V_1}{K_e\phi} - \frac{R_a T}{K_e K_t \phi^2} = \omega_{10} - \Delta\omega \tag{3.30}$$

and
$$\omega_2 = \frac{V_2}{K_e\phi} - \frac{R_a T}{K_e K_t \phi^2} = \omega_{20} - \Delta\omega \tag{3.31}$$

In the case of a shunt motor, it is seen that the no-load speeds vary in proportion to applied voltages, i.e , $\frac{\omega_{10}}{\omega_{20}} = \frac{V_1}{V_2}$, and the drop in speed $\Delta\omega$ remains the same for a specified load torque. Hence, the family of speed-torque characteristics of a shunt motor (and separately excited motor) for different armature-terminal voltages will be as shown in Fig. 3.26.

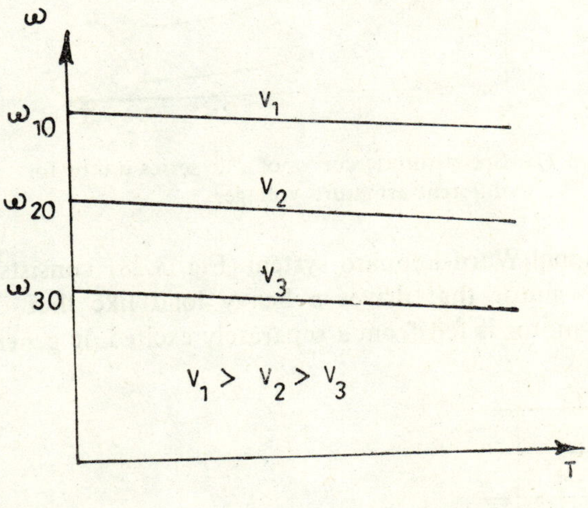

Fig. 3.26. Speed-torque curves of a dc shunt motor for different armature voltages.

The set of speed-torque characteristics corresponding to series motors (Fig. 3.27) can also be deduced using the above equations. Of course, it must be noted that irrespective of the applied voltage to the motor, the no load speed is infinite.

Normally, this method gives speeds below rated, since voltage magnitudes greater than rated voltage should not be impressed on the motor. But, by varying the field current (applied voltage to the motor remaining at its rated value) of separately excited or shunt motors, speeds above the rated value can be obtained.

In order to provide controlled armature dc voltage from the readily available ac power, an ac motor-dc generator set is used. The motor-generator set known as Ward-Leonard set was first used about eighty years ago and remains as the basis for many modern drive systems.

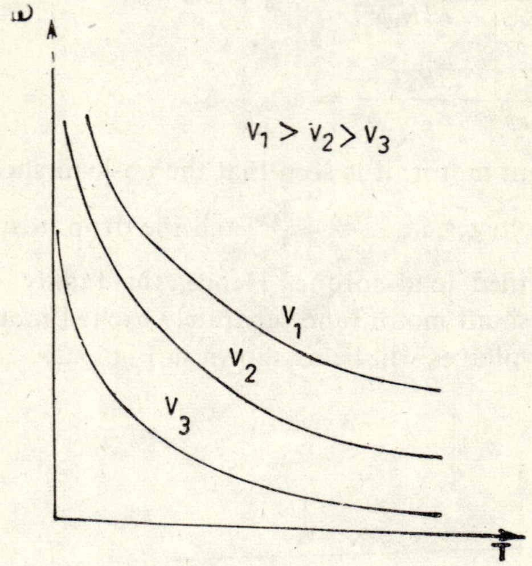

Fig. 3.27 Speed-torque curves of a dc series motor for different armature voltages.

The conventional Ward-Leonard system (Fig. 3.28) consists of a separately excited dc motor that drives a heavy load like that offered by a rolling mill. The motor is fed from a separately excited dc generator, which

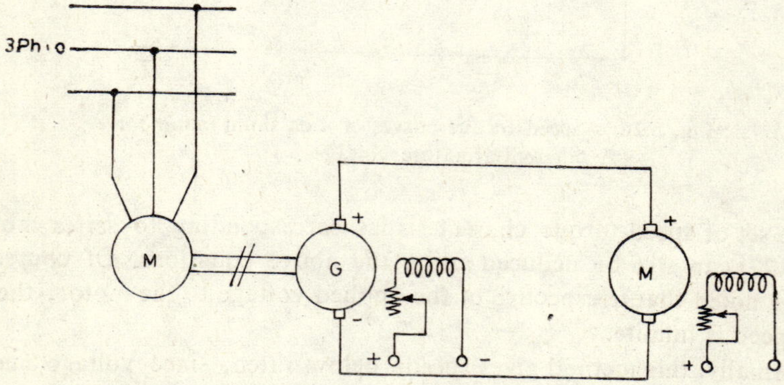

Fig. 3.28. Schematic diagram of Ward-Leonard system.

is driven either by a squirrel cage induction motor (for low powers) or a wound-rotor induction motor or synchronous motor (for high powers). Reversal of direction of rotation of the motor is achieved by reversing the polarity of the field winding of the dc generator by means of switches.

The steady state speed-torque characteristics of a Ward-Leonard system can be derived from the equivalent circuit shown in Fig. 3.29. R_a represents the total resistance of the armatures of the two machines. Under normal conditions of operation the emf induced in the generator E_g will be slightly larger than the emf induced in the motor E_m.

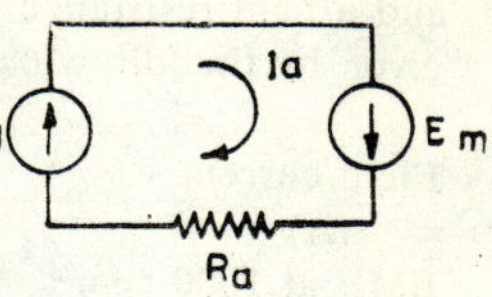

Fig. 3.29. Equivalent circuit.

It follows from Fig. 3.29 that

$$E_g + E_m = I_a R_a \tag{3.32}$$

$$E_g = K_g \phi_g \omega_g \quad \text{and} \quad E_m = -K_m \phi_m \omega_m \tag{3.33}$$

The electromagnetic torque of the motor is given by

$$T = K_m \phi_m I_a \tag{3.34}$$

By eliminating E_g, E_m, and I_a from equations (3.32) to (3.34), the motor speed is given by

$$\omega_m = \frac{K_g \phi_g \omega}{K_m \phi_m} - \frac{R_a T}{K_m \phi_m^2} \tag{3.35}$$

It will be seen that when the motor forms part of a Ward-Leonard system, the speed drop is larger than it was when the motor is fed from constant voltage mains, since R_a is greater than the resistance of motor armature.

The speed-torque characteristics obtained using Eqn. (3.35) depict two areas of interest, as illustrated in Fig. 3.30. The first corresponds to motor

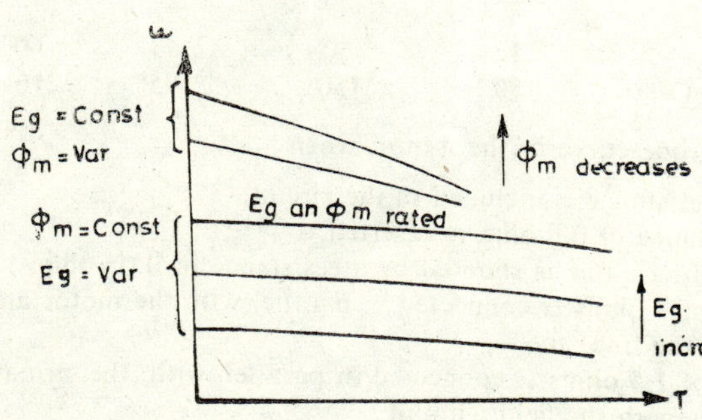

Fig. 3.30. Speed-torque characteristics of a Ward-Leonard system.

operation at variable voltage supply and rated excitation. They form a family of parallel straight lines slightly drooping towards x-axis. The second area corresponds to motor operation at rated supply voltage and reduced field excitation. The characteristics are drooping straight lines of different slope.

PROBLEMS

1. A 220 V, 21 A, 1000 rpm dc shunt motor has an armature resistance of 0.05 ohm and a field resistance of 220 ohms. The magnetization curve for the machine is given by the following table:

Field current (A)	0.2	0.4	0.6	0.8	1.0	1.2	1.4
EMF at 1000 rpm (V)	50	100	150	190	219	235	245

Plot the speed-torque curve for the motor when

(a) no external resistances are included,
(b) a resistance of 0.05 ohm is connected in series with the armature, and
(c) a resistance of 110 ohms is included in series with the field winding.

2. The motor in Problem 1 drives a certain load at 1000 rpm taking a current of 21 A from the line. If it is desired to raise its speed to 1500 rpm, what value of resistance must be included in the field circuit when the load is of such a nature that (a) the electromagnetic torque required of the motor remains constant, regardless of speed and (b) the electromagnetic power required of the motor remains constant, regardless of the speed.

3. The motor in Problem 1 draws an armature current of 21 A on full load and runs at 1000 rpm. If a resistance of 0.5 ohm is inserted in series with armature, determine

(a) the speed at full load torque,
(b) the speed at 60 per cent of full load torque, and
(c) the ratio of starting torque to full load torque.

4. A 220 V dc series motor has an armature resistance of 0.05 ohm and a field resistance of 0.05 ohm. While running as a generator at 1400 rpm, it gave the following results:

Current (A)	15	30	45	60	75
Terminal voltage (V)	80	150	195	216	240

Plot the speed-torque curve of the motor, when

(a) no external resistance is included in the circuit,
(b) a series resistance of 0.6 ohm is inserted,
(c) the field winding alone is shunted by a resistance of 0.05 ohm,
(d) a resistance of 5 ohms is connected in parallel with the motor and 1.5 ohms in series with the line, and
(e) a resistance of 1.5 ohms is connected in parallel with the armature and 2 ohms in series with the field winding.

5. For the motor in Problem 4, determine the values of the resistances R_D in parallel with the armature and R in series with the field for the shunted armature connection to give a speed-torque characteristic with (i) a speed of 500 rpm at zero torque and (ii) a torque of 100 N-m at zero speed.

6. Plot the speed-torque curves for the motor in Problem 4 when connected as a shunt motor as shown in Fig. 3.31 with

(a) $R_1 = 2.5$ ohms, $R = 1$ ohm, and
(b) $R_1 = 2.5$ ohms, $R = 2.5$ ohms.

7. If a series motor were connected as shown in Fig. 3.32. Show that the starting torque has a maximum value, when

$$R_2 = \frac{R(R_1 + R_f)}{R + R_1 + R_f} - R_a.$$

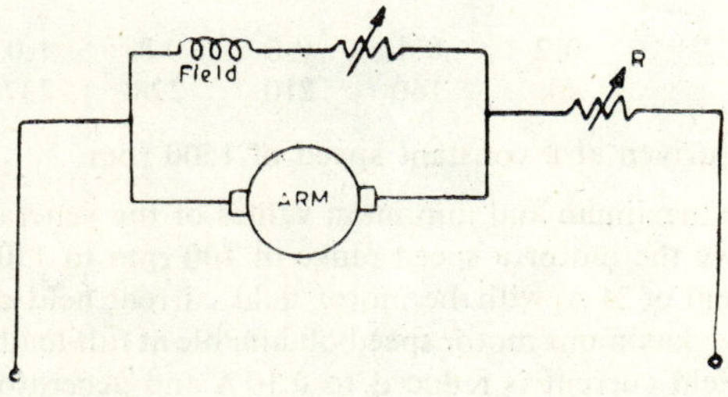

Fig. 3.31. Motor connections for Problem 6.

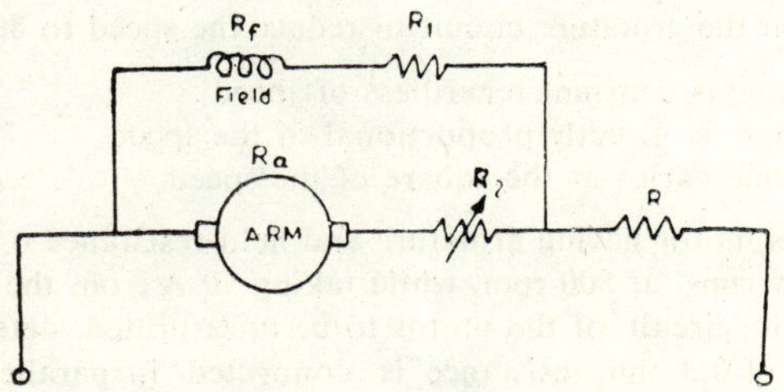

Fig. 3.32. Motor connections for Problem 7.

8. A 230 V dc series motor used in lifts has a resistance of 0.2 ohm. At a speed of 1800 rpm it takes 50A. Determine the resistance to be added in series with the motor
   (a) to limit the speed to 3600 rpm when the current is 12.5 A, assuming that the magnetization curve is a straight line between zero and 50 A;
   (b) to make the speed 900 rpm when the current is 70 A, taking the flux per pole at 70 A is 20 per cent greater than that at 50 A.

   At what speed will the motor run, when directly connected to the 230 V mains and drawing a current of 70 A?

9. A 230 V dc series motor having the magnetization curve given below, runs at 750 rpm with a certain load and takes a current of 60 A from the lines:

Current (A)	10	20	30	40	50	60	70	80	90
Flux per pole (mWb)	2.9	5.5	7.4	8.5	9.2	9.7	10.1	10.6	10.8

The resistances of the field and armature are 0.05 and 0.1 ohm respectively. Determine the value of the resistance of the divertor connected in parallel with the field in order to increase the speed to 900 rpm, the load torque remaining the same as before.

(**Hint:** Adopt trial and error method. Assume a lower value of flux corresponding to increased speed. Then, both new values of I_a and E are known, from which the divertor resistance can be determined. Then, new value of field current can be found. Check whether this field current corresponds to the assumed value of flux.)

10. Two identical 5 kW, 220 V, 24 A dc shunt machines are to be used as the generator and motor in a Ward-Leonard system. The armature resistance of each machine is 0.5 ohm and the magnetization curve for each machine at 1500 rpm is given below:

Field current, A	0.2	0.4	0.6	0.8	1.0	1.2	1.4
Induced emf, V	80	160	210	228	237	243	247

The generator is driven at a constant speed of 1500 rpm.

(a) Determine the maximum and minimum values of the generator field current required to give the motor a speed range of 100 rpm to 1500 rpm at full load armature current of 24 A, with the motor field current held constant at 0.6 A.
(b) Determine the maximum motor speed obtainable at full load armature current if the motor field current is reduced to 0.10 A and generator field current is not allowed to exceed 1.2 A.

11. A 250 V, 10 kW, 1200 rpm dc shunt motor has a full load efficiency of 80 per cent. Its field and armature resistances are 110 ohms and 0.25 ohm respectively. Calculate the value of the resistance to be inserted in series with the armature and the power lost in the armature circuit to reduce the speed to 80 per cent when

(a) the load torque is constant regardless of speed,
(b) the load torque is directly proportional to the speed,
(c) the load torque varies as the square of the speed.

12. A 250 V dc series motor having armature and field resistance 0.25 ohm and 0.15 ohm respectively runs, at 800 rpm, while taking 50 A from the supply. Assuming the magnetic circuit of the motor to be unsaturated, determine its speed, when a divertor of 0.3 ohm resistance is connected in parallel with the field winding of the motor, which is subjected to a load torque whose torque

(a) remains constant regardless of speed,
(b) varies in direct proportion to the speed,
(c) varies as the square of the speed.

13. A dc series motor on load operating at 250 V dc mains draws 25 A and runs at 1200 rpm. Armature and field resistances are 0.1 ohm and 0.3 ohm respectively. A resistance of 25 ohms is placed in parallel with the armature of the motor. Determine:

(a) The speed of the motor with the shunted armature connection, if the magnetic circuit remains unsaturated and the load torque remains constant, and
(b) the no load speed of the motor with the shunted armature connection.

14. An adjustable speed dc shunt motor has a speed range of 3:1. If the motor draws an armature current of 50 A at 1000 rpm, calculate the current at the speed of 3000 rpm when speed control is achieved by (i) field flux control, and (ii) armature voltage control, with the motor driving a load whose,

(a) load torque is constant, and
(b) load power is constant.

Neglect all losses and effect of armature reaction.

Answers

2. (a) 156.7 ohms (b) 156.7 ohms
3. (a) 951.82 rpm (b) 972.92 rpm (c) 19.48
5. (i) For $I_f = 60$ A, $R_D = 1.286$ ohms and $R = 2.3$ ohms.
   (ii) For $I_f = 75$ A, $R_D = 0.219$ ohms and $R = 2.66$ ohms.
8. (a) 9.4 ohms (b) 1.2 ohms, 1472.73 rpm
9. 0.05 ohm
10. (a) 0.98 A and 0.08 A (b) 8212.5 rpm
11. (a) 1 ohm and 2.847 kW (b) 1.31 ohms and 2.274 kW
    (c) 1.70 ohms and 1.818 kW
12. (a) 973 rpm (b) 912 rpm (c) 879 rpm
13. (a) 986.37 rpm (b) 3125.38 rpm
14. (a) (i) 150 A; (ii) 50 A (b) (i) 50 A; (ii) 50/3 A

Chapter 4

Characteristics of ac Motors

4.1 Three Phase Induction Motors

The speed versus torque characteristics of induction motors are quite important in selecting an induction motor drive. In addition, the ratio of maximum torque to rated torque, ratio of starting current to rated current, ratio of starting torque to rated torque and the ratio of no load current to rated current are of equal significance. The most convenient method of determining the above characteristics is by means of the equivalent circuit of the induction motor.

4.1.1 Steady State Characteristics

Figure 4.1 shows the approximate equivalent circuit, which represents the induction motor performance at a slip s, with sufficient accuracy. All quantities, defined in this figure are per phase values and all rotor quantities have been referred to stator. The term R_2/s may be considered as the sum of two resistances R_2 and $\frac{(1-s)R_2}{s}$, where R_2 is the effective rotor resistance and $\frac{(1-s)R_2}{s}$ is a fictitious resistance which represents, electrically, the mechanical power developed in the rotor.

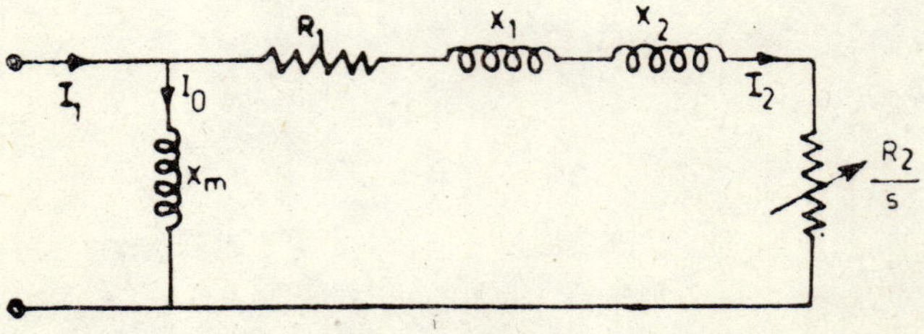

Fig. 4.1. Approximate equivalent circuit of induction motor.

From the equivalent circuit, it is obvious that

(i) input power to the rotor $= 3I_2^2 \frac{R_2}{s}$ watts

(ii) rotor copper loss $= 3I_2^2R_2$ watts and

(iii) mechanical power developed $= 3I_2^2 \dfrac{R_2(1-s)}{s}$ watts.

The developed power may also be expressed as the product of torque developed and actual speed of rotation of rotor, i.e., $P = T\cdot\omega_r = T\cdot\omega_s(1-s)$ where, ω_s is the synchronous speed in mechanical rad./sec. and T the torque in N-m.

Thus,
$$T = \frac{3}{\omega_s} I_2^2 \frac{R_2}{s} \text{ Newton metres} \tag{4.1}$$

But
$$I_2 = \frac{V_1}{\sqrt{(R_1 + R_2/s)^2 + (X_1 + X_2)^2}} \tag{4.2}$$

Therefore,
$$T = \frac{3}{\omega_s} \cdot \frac{V_1^2}{(R_1 + R_2/s)^2 + X^2} \cdot \frac{R_2}{s} \text{ Newton metres,} \tag{4.3}$$

where $X = (X_1 + X_2)$.

Typical speed-torque curves are shown in Fig. 4.2. The two curves are identical; one being drawn for the forward direction of rotation and the other for the reverse. The fact that the motoring curve in the first quadrant extends into the fourth quadrant indicates that positive torque is developed even though the motor is rotating in the reverse direction. This same curve extends into the second quadrant indicating a negative torque with speeds above synchronous. This regenerative action is possible only if the system to which the motor is connected can supply the required reactive power for

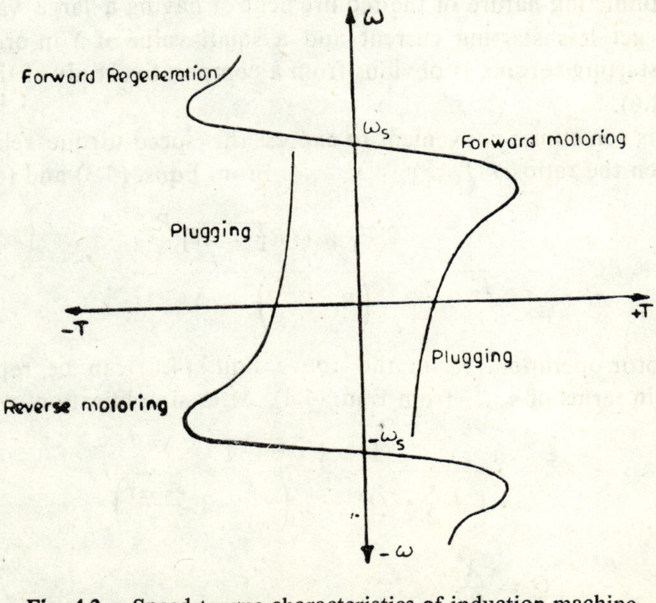

Fig. 4.2. Speed-torque characteristics of induction machine.

excitation. The maximum torque may be found by first determining the slip at which the maximum torque or pull out occurs. Differentiating Eqn. (4.3) with respect to s and equating it to zero for a maximum, we get

$$s_{\max T} = \pm \frac{R_2}{\sqrt{R_1^2 + X^2}} \tag{4.4}$$

Plus and minus sign in the above equation corresponds to motoring and generating operation respectively. Substituting Eqn. (4.4) in Eqn. (4.3), the maximum torque will be obtained as.

$$T_{\max} = \frac{3}{\omega_s} \cdot \frac{V_1^2}{2[R_1 \pm \sqrt{R_1^2 + X^2}]} \text{ Newton metres} \tag{4.5}$$

From the above equations, it is clear that the magnitude of the maximum torque is independent of R_2 although the slip at which the maximum occurs is a function of R_2. Also, $T_{\max}$ developed in the generating mode is greater than that produced in the motoring mode, since

$$(R_1 - \sqrt{R_1^2 + X^2}) < (R_1 + \sqrt{R_1^2 + X^2}).$$

The starting torque T_{st} is obtained from Eqn. (4.3) by putting s equal to 1.

$$T_{st} = \frac{3}{\omega_s} \cdot \frac{V_1^2 R_2}{(R_1 + R_2)^2 + X^2} \text{ Newton metres} \tag{4.6}$$

Similarly, the starting current is found from Eqn. (4.2) by substituting $s = 1$.

$$I_{2st} = \frac{V_1}{\sqrt{(R_1 + R_2)^2 + X^2}} \text{ amperes} \tag{4.7}$$

The conflicting nature of the requirement of having a large value of X so as to get less starting current and a small value of X in order to have a large starting torque, is obvious from a comparative study of the Eqns. (4.7) and (4.6).

It is sometimes convenient to express the speed-torque relation as one between the ratios $T/T_{\max}$ and $s/s_{\max T}$. From Eqns. (4.3) and (4.5), we have

$$\frac{T}{T_{\max}} = \frac{2[R_1 + \sqrt{R_1^2 + X^2}]\,\dfrac{R_2}{s}}{\left(R_1 + \dfrac{R_2}{s}\right)^2 + X^2} \tag{4.8}$$

for motor operation. R_2 in the above Eqn. (4.8) can be replaced by its value in terms of $s_{\max T}$ from Eqn. (4.4). After simplification, we get

$$\frac{T}{T_{\max}} = \frac{1 + \sqrt{Q^2 + 1}}{1 + \dfrac{1}{2}\sqrt{Q^2 + 1}\left(\dfrac{s}{s_{\max T}} + \dfrac{s_{\max T}}{s}\right)}, \tag{4.9}$$

where $\quad Q = \dfrac{X}{R_1}.$

A more simple but approximate expression for the torque slip relation can be obtained by substituting $Q = \infty$ in Eqn. 4.9. This is equivalent to saying that the effect of stator resistance R_1 is negligible. Under this condition, Eqn. (4.9) becomes,

$$\frac{T}{T_{max}} = \frac{2}{s/s_{maxT} + s_{maxT}/s} \tag{4.10}$$

From Eqn. (4.10) it is obvious that if the maximum torque and the slip at which it occurs are specified, the speed-torque characteristic is approximately fixed throughout the entire speed range. This statement is subject to the condition that the motor parameters are constant and therefore is not applicable to motors with variable rotor resistance.

Example 4-1. A 3-phase induction motor, at rated voltage and frequency, has a maximum torque of 225 per cent and a starting torque of 150 per cent of full load torque. Neglect stator resistance and rotational losses and assume constant rotor resistance. Calculate (a) the slip at maximum torque and (b) the slip at full load.

Solution: (a) We know that $\dfrac{T_{max}}{T} = \dfrac{\dfrac{s_{maxT}}{s} + \dfrac{s}{s_{maxT}}}{2}$

At starting $\quad s = 1.$

Therefore, $\quad \dfrac{T_{max}}{T_{st}} = \dfrac{\dfrac{s_{maxT}}{1} + \dfrac{1}{s_{maxT}}}{2}$

i.e., $\quad \dfrac{2.25}{1.5} = \dfrac{s_{maxT} + \dfrac{1}{s_{maxT}}}{2}$

i.e., $\quad s_{maxT} + \dfrac{1}{s_{maxT}} = 3$

i.e., $\quad s^2_{maxT} - 3\, s_{maxT} + 1 = 0$

Hence, $\quad s_{maxT} = \dfrac{3 \pm \sqrt{9-4}}{2} = \dfrac{3 \pm \sqrt{5}}{2}$

$\quad = 2.618$ or 0.382

Slip greater than unity will indicate braking operation and hence $s_{maxT} = 0.382$ for motor operation.

(b) $\quad \dfrac{T_{max}}{T_f} = 2.25 = \dfrac{\dfrac{0.382}{s_f} + \dfrac{s_f}{0.382}}{2}$

Hence, $\quad s_f^2 - 4.5 \times 0\,382\, s_f + (0.382)^2 = 0$

i.e., $\quad s_f^2 - 1.719\, s_f + 0.146 = 0$

Therefore
$$s_t = \frac{1.719 \pm \sqrt{2.955 - 0.584}}{2}$$
$$= \frac{1.179 \pm 1.539}{2}$$
$= 0.09$; the other value greater than unity is neglected.

4.1.2 No Load Current of Induction Motors

The no load current of an induction motor is yet another factor which determines its performance. It is one of those important data required for constructing the circle diagram from which the performance of the motor can be predicted. Unfortunately, the value of I_0 is neither included in specification nor given on the name plate. Of course, it can be experimentally determined simply if the motor were either not connected with any equipment or can be disconnected easily from it. The magnitude of the no load current obtained experimentally will be larger than the actual value, if the equipment to which the motor is attached cannot be disconnected.

The value of the no load current more or less depends on the magnetizing current required. Usually the magnitnde of I_0 varies from 20 to 60 per cent of the rated current. It is large in case of totally enclosed motors and motors having larger air gaps (As those used in cranes). Other conditions remaining the same, the ratio of no load current to rated current I_0/I_{fl} increases, as the rated speeds of the motors decrease. Typical values of the ratio of I_0/I_{fl} are given in Table 4.1.

Table 4.1 Approximate values of the no load current of induction motors in percentage

Power output kW	$I_0/I_{fl} \times 100$ for different, rated synchronous speeds of induction motors in rpm					
	3000	1500	1000	750	600	500
1-5	45	65	70	75	80	85
5-10	40	60	65	70	75	80
10-25	30	55	60	65	70	75
25-50	20	50	55	60	65	70
50-100	—	40	45	50	55	60

4.1.3 Relationship Between the Ratio of Starting Current to Full Load Current and the Ratio of Starting Torque to Full Load Torque

From Eqn. 4.1, we have

$$T_{st} = 3\, I_{2\,st}^2 R_{2st}/\omega_s \tag{4.11}$$

and

$$T_{fl} = 3\, I_{2\,sf}^2 R_{2sf}/s_f\omega_s \tag{4.12}$$

Note that R_{2st} will be higher than R_{2sf} due to the significant difference in the frequency of rotor currents at starting and at full load.

Hence
$$\frac{T_{st}}{T_f} = \left(\frac{I_{2\,st}}{I_{2f}}\right)^2 \left(\frac{R_{2\,st}}{R_{2f}}\right) s_f \tag{4.13}$$

From the circle diagram of the induction motor it will be clear that the ratio of rotor currents referred to stator corresponding to starting and full load do not differ very much from the ratio of stator currents themselves and can be approximately taken as

$$\left(\frac{I_{2st}}{I_{2f}}\right) = 0.9 \left(\frac{I_{1st}}{I_{1f}}\right) \tag{4.14}$$

Taking $R_{2st} = 1.2\, R_{2\,f}$ and substituting these values in Eqn. (4.13), we get

$$\left(\frac{T_{st}}{T_f}\right) = \left(\frac{I_{1st}}{I_{1f}}\right)^2 s_f \tag{4.15}$$

The above expression is valid only for normal induction motors and not for motors having specially designed rotors.

Example 4-2. For a three phase induction motor with negligible stator resistance and no load current, show that the ratio of the starting current to its stator current at any slip can be expressed as

$$\frac{I_{st}}{I} = \sqrt{\frac{s^2 + s_{\max T}^2}{s^2(1 + s_{\max T}^2)}}$$

where I_{st} = starting current,

I = stator input current at any slip s and

$s_{\max T}$ = slip at which maximum torque occurs.

Solution: With negligible stator resistance and no load current, the stator current of an induction motor at any slip s can be expressed as

$$I = \frac{V_1}{\sqrt{\left(\frac{R_2}{s}\right)^2 + X^2}}$$

$$I_{st} = \frac{V_1}{\sqrt{(R_2)^2 + X^2}}$$

Therefore,

$$\frac{I_{st}}{I} = \sqrt{\frac{\left(\frac{R_2}{s}\right)^2 + X^2}{R_2^2 + X^2}} \tag{4.16}$$

With negligible stator resistance, we have

$$s_{\max T} = \frac{R_2}{X}$$

i.e.
$$X = \frac{R_2}{s_{\max T}}$$

Substituting this value of X in Eqn. (4.16), we get

$$\frac{I_{st}}{I} = \sqrt{\frac{(R_2/s)^2 + (R_2/s_{maxT})^2}{R_2^2 + (R_2/s_{maxT})^2}}$$

$$= \sqrt{\frac{s^2 + s_{maxT}^2}{s^3(1 + s_{maxT}^2)}}$$

4.1.4 Modified Speed-Torque Characteristics of Three Phase Induction Motors

There are, in general, five methods of modifying the speed-torque characteristics of three phase induction motors: (i) Variation of applied voltage (ii) Variation of supply frequency (iii) Introduction of balanced resistances or inductances in the stator circuit (iv) Addition of balanced resistors in the rotor circuit and (v) Injection of voltages in the rotor circuit. Obviously, the last two methods are applicable only to slip ring induction motors.

(i) *Variation of applied voltage*: The torque at any value of slip varies as the square of the applied voltage as indicated by Eqn. (4.3); using this property a family of speed torque curves as shown in Fig. 4.3 can be computed for the machine when it operates at different voltages.

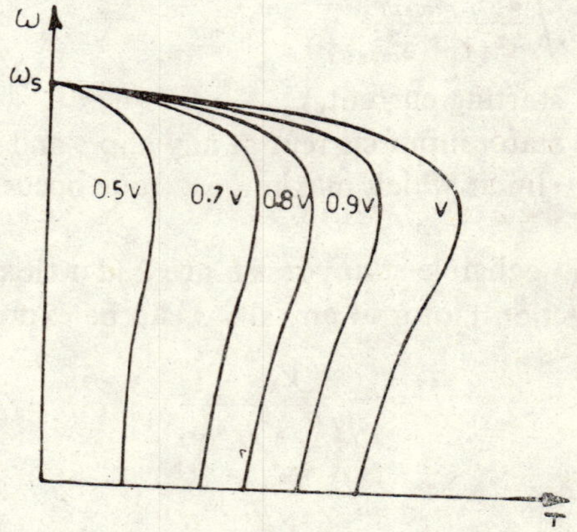

Fig. 4.3. Speed-torque curves with stator voltage control.

The curves of Fig. 4.3 indicate that the slip at maximum torque is independent of the terminal voltage; this fact is confirmed by Eqn. (4.4). The range of speeds within which steady state operation (for constant torque loads) may take place is the same for all voltages, namely between the speed corresponding to maximum torque and synchronous speed. Within that region, there will be a small drop in speed with decrease in voltage, but the range of speeds obtainable in this manner is quite small.

Example 4-3. Assuming steady state stability of the drive system, derive an expression for the speed of the induction motor driving a load torque proportional to some power of the speed, when the supply voltage falls by a certain percentage.

If an induction motor having a rated slip of 4 per cent and $s_{\max T}$ of 20 per cent drives a constant load torque, determine the slip at which the motor will run, when the supply voltage falls by 20 per cent.

Solution: The torque developed by the motor at the rated voltage can be obtained from Eqn. (4.10) as,

$$T = \frac{2T_{\max}\, s_{\max T}\, s}{s^2 + s_{\max T}^2} \tag{4.17}$$

Assume that the applied voltage is reduced to x times the rated voltage, where x is a fraction.

Torque at this reduced voltage will be

$$T_x = T x^2 = x^2 \frac{2\, T_{\max}\, s_{\max T}\, s}{s^2 + s_{\max T}^2} \tag{4.18}$$

The load torque T_L is given as proportional to (speed)^n, where 'n' is any number, depending on the nature of the load.

Hence

$$T_L = K\,(\text{speed})^n = K\,(1 - s)^n, \tag{4.19}$$

where K is a constant.

Denoting the quantities corresponding to the new applied voltage with a dash above, the steady state condition gives us

$$T = T_L \text{ and } T' = T_L' \tag{4.20}$$

Using Eqns. (4.17 – 4.20), we get

$$\frac{2\, T_{\max} \cdot s_{\max T}\, s}{s^2 + s_{\max T}^2} = K\,(1 - s)^n \text{ and}$$

$$\frac{2T_{\max} \cdot s_{\max T}\, s' \cdot x^2}{s'^2 + s_{\max T}^2} = K\,(1 - s')^n.$$

Dividing one by the other, we have

$$\frac{x^2 \cdot s'}{(1 - s')^n\,(s'^2 + s_{\max T}^2)} = \frac{s}{(1 - s)^n\,(s^2 + s_{\max T}^2)}$$

This is the expression, which determines the new speed.

Given that $s = 0.04$; $s_{\max T} = 0.20$, $x = (1 - 0.2) = 0.8$ and $n = 0$, since the load torque is independent of the speed.

For a constant torque load, the expression derived above becomes,

$$\frac{x^2 s'}{s'^2 + s_{\max T}^2} = \frac{s}{s^2 + s_{\max T}^2}$$

Substituting the given values for x, s and s_{maxT}, we get

$$\frac{0.64\, s'}{s'^2 + 0.04} = \frac{0.04}{0.0016 + 0.04}$$

i.e. $$0.04\, s'^2 - 0.0266\, s' + 0.0016 = 0$$

$$s' = \frac{0.0266 \pm \sqrt{0.000707 - .000256}}{0.08}$$

$$= \frac{0.0266 \pm 0.0212}{0.08}$$

$$= 0.5975 \text{ or } 0.0675$$

$s' = 0.0675$, the other value being greater than s_{maxT} will not give a steady point of operation.

(ii) *Variation of supply frequency*: A change in the frequency of the power supply for an induction motor will result in a corresponding change in the synchronous speed and some change in the motor characteristics. In order to maintain the air gap flux at its normal value, it is necessary to keep E_1/f constant. Since controlling the induced emf E_1 is difficult, it is customary to vary the magnitude of the applied voltage in the same ratio of the frequency, thus keeping V_1/f constant and hence the flux approximately constant.

Let $\alpha = \frac{f}{f_{max}} < 1$, where f is the operating frequency and f_{max} is the maximum possible frequency of the supply to which the motor may be subjected. Usually, f_{max} corresponds to the rated frequency of the motor. The synchronous speed becomes $\alpha\, \omega_s$, the applied voltage is αV, and all the reactances are αX. Substituting these in the torque equation (Eqn. 4.3) and simplifying, we have

$$T = \frac{1}{\omega_s} \cdot \frac{3V_1^2\, R_2/\alpha s}{\left(\frac{R_1}{\alpha} + \frac{R_2}{\alpha s}\right)^2 + X^2} \tag{4.21}$$

It may be seen that this expression is of the same form as the original torque equation, but all resistances have become larger by the factor $1/\alpha$. Similar results can be obtained with the expressions for starting and maximum torque.

$$T_{st} = \frac{3}{\omega_s} \cdot \frac{V_1^2 \left(\frac{R_2}{\alpha}\right)}{\left(\frac{R_1}{\alpha} + \frac{R_2}{\alpha}\right)^2 + X^2} \tag{4.22}$$

$$T_{max} = \frac{3}{\omega_s} \cdot \frac{V_1^2}{2\left[\frac{R_1}{\alpha} \pm \sqrt{\left(\left(\frac{R_1}{\alpha}\right)^2 + X^2\right)}\right]} \tag{4.23}$$

The slip at which maximum torque occurs becomes

$$s_{maxT} = \pm \frac{\left(\frac{R_2}{\alpha}\right)}{\sqrt{\left(\frac{R_1}{\alpha}\right)^2 + X^2}} \tag{4.24}$$

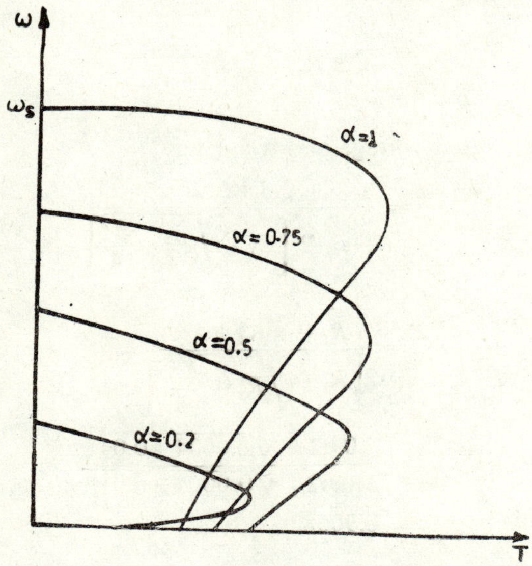

Fig. 4.4. Speed-torque curves with V/f control.

Typical speed-torque curves for four different frequencies are shown in Fig. 4.4. The slip at which maximum torque occurs becomes larger as the operating frequency decreases and the maximum torque gets reduced slightly. The starting torque increases for small reductions in frequency, but attains a maximum and then decreases with further reduction in frequency.

The reduction in developed torque at low frequencies is partly due to the apparent increase in resistances of the machine and also due to the decrease in air gap flux, which the above torque expression (Eqn. 4.21) neglects. This reduction in air gap flux, which arises from the voltage drop in the stator impedance, will obviously depend on the input current and hence on the load on the motor. It is possible to compensate for this change in air gap flux by making the ratio V_1/f to increase as α becomes small. However, an increase in V_1/f so as to keep the air gap flux at normal value, with the machine loaded will result in increased magnetised current being drawn from the supply.

Example 4-4. A 3-phase, 50 Hz induction motor, represented by the equivalent circuit constants $X_1 = X_2 = 0.1$ ohm and $R_1 = R_2 = 0.02$ ohm is to be operated at one half of its rated voltage and 25 Hz frequency. Calculate

(a) the maximum torque at this reduced voltage and frequency operation in terms of its normal value, and

(b) the starting torque at this reduced frequency and voltage in terms of its normal value.

Solution:

(a) T_{max} at rated voltage and frequency

$$= \frac{3}{\omega_s} \frac{V_1^2}{2[R_1 + \sqrt{R_1^2 + X^2}]}$$

where $X = (X_1 + X_2)$

T_{max} at one-half rated voltage and frequency

$$= \frac{3\, V_1^2 . 2}{4 \cdot \omega_s \cdot 2 \left[R_1 + \sqrt{R_1^2 + \frac{X^2}{4}} \right]}$$

$$\text{Therefore, } \frac{T_{max} \text{ at } \frac{V}{2}, \frac{f}{2}}{T_{max} \text{ at } V, f} = \frac{R_1 + \sqrt{R_1^2 + X^2}}{2\left[R_1 + \sqrt{R_1^2 + \frac{X^2}{4}} \right]}$$

$$= \frac{0.02 + \sqrt{0.0004 + 0.04}}{2[0.02 + \sqrt{0.0004 + 0.01}]}$$

$$= 0.9058.$$

(b) T_{st} at rated voltage and frequency

$$= \frac{3 V_1^2 R_2}{\omega_s \,[(R_1 + R_2)^2 + X^2]}$$

T_{st} at one-half rated voltage and frequency

$$= \frac{3 V_1^2 \times 2 R_2}{4 \omega_s \,[(R_1 + R_2)^2 + (X^2/4)]}$$

$$\text{Therefore, } \frac{T_{st} \text{ at } \frac{V}{2}, \frac{f}{2}}{T_{st} \text{ at } V, f} = \frac{(R_1 + R_2)^2 + X^2}{2\left[(R_1 + R_2)^2 + \frac{X^2}{4} \right]}$$

$$= \frac{(0.04)^2 + (0.2)^2}{2\left[(0.04)^2 + \frac{(0.2)^2}{4} \right]}$$

$$= 1.793.$$

(iii) *Introduction of stator impedance*: Balanced resistors or inductors can be added to the stator circuit so as to reduce the voltage at the machine terminals. Under these conditions, the motor terminal voltage becomes a function of the motor current so that this voltage changes as the motor accelerates. Typical speed torque curves are shown in Fig. 4.5 for the cases of added resistance and inductance.

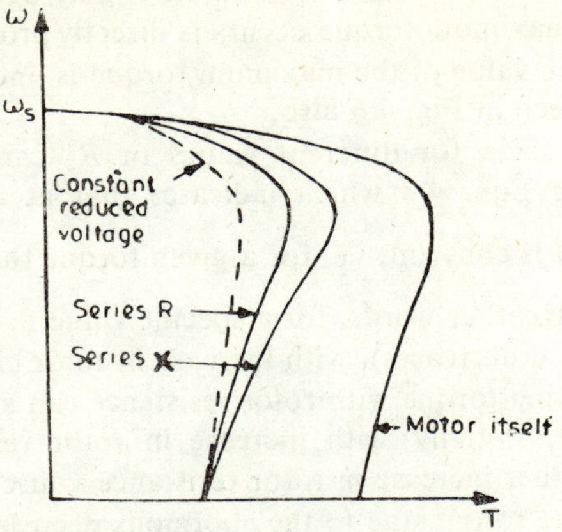

Fig. 4.5. Speed-torque curves with stator impedance control.

If the additional resistance (R) or inductance (X) were chosen such as to give the same starting torque, the speed torque characteristic corresponding to additional inductance would have larger torques than with additional resistance. Besides, both these characteristics enable us to get larger torques than with the characteristic obtained with reduced applied voltage, which gives the same starting torque.

Modifying the characteristic by means of introducing external resistance in the stator circuit will improve the power factor, but at the expense of slightly greater losses at starting. These losses are minimized with reactor starting, but the power factor becomes poor.

(iv) *Addition of balanced resistors in the rotor circuit*: The introduction of external resistances in the rotor circuit of a slip ring induction motor will modify the speed torque characteristics as shown in Fig. 4.6. This is a method by which any operating speed between zero and synchronous

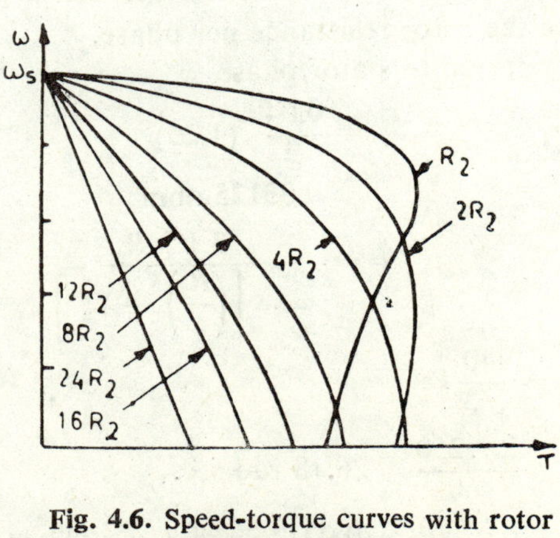

Fig. 4.6. Speed-torque curves with rotor resistance control.

speed can be obtained. From Eqns. 4.4 and 4.5 it may be observed that the speed at which the maximum torque occurs is directly proportional to rotor resistance R_2, but the value of the maximum torque is independent of R_2. These facts can be seen in Fig. 4.6 also.

The family of curves for different values of R_2 can be plotted quite accurately by noting Eqn. 4.3 which indicates that at a given value of torque the ratio $\frac{R_2}{s}$ is constant, i.e. for a given torque the slip is directly proportional to R_2. In other words, for a specific value of torque, the slip increases (or the speed decreases), with increase in rotor circuit resistance.

Variation of starting torque with rotor resistance can also be seen from these curves. While, initially with increase in rotor resistanee, starting torque increases, further increase in rotor resistance causes a decrease in the value of starting torque due to the enormous decrease in the value of the starting current.

Example 4-5. A 3-phase, 440 V, 50 Hz, 110 kW, 24 pole, 245 rpm slip ring induction motor has both its stator and rotor windings connected in star. The ratio of stator to rotor turns is 1.25. The resistance measured between each pair of slip rings is 0.04 ohm. This motor drives a fan which requires 110 kW at the full load speed of the motor. The torque required to drive the fan varies as the square of the speed. What resistances should be connected in series with each slipring so that the fan will run at 175 rpm? Neglect stator resistance, stator leakage reactance and rotational losses of the motor.

Solution: Synchronous speed $= \frac{120.50}{24} = 250$ rpm

Full load slip $= \frac{250-245}{250} = 0.02$

Torque at full load $= \frac{110.1000.60}{2\pi.245}$

$= 4287.44$ N-m

Since the rotor is star connected, the resistance between a pair of sliprings will be equal to twice the rotor resistance per phase.

Rotor resistance referred to stator/phase

$$= \frac{0.04}{2} \cdot (1.25)^2$$

$$= 0.03125 \text{ ohm}$$

$$T = \frac{3 V_1^2 R_2}{\omega_s . s \left[\left(\frac{R_2}{s}\right)^2 + X_2^2\right]}$$

Substituting $V_1 = \frac{440}{\sqrt{3}} V,$

$$\omega_s = \frac{2\pi.250}{60} = 26.18 \text{ rad/sec.}$$

$s = 0.02$, $R_2 = 0.03125$ and $T = 4287.44$,

we get,

$$4287.44 = \frac{(440)^2.0.03125}{26.18\cdot 0.02\left[\left(\frac{0.03125}{0.02}\right)^2 + X_2^2\right]}$$

$$X_2^2 = 0.2535 \text{ and } X_2 = 0.5035$$

$$\text{Slip at 175 rpm} = \frac{250 - 175}{250} = 0.3$$

Let the rotor resistance referred to stator/phase corresponding to the above slip be R_2'. Representing the torque at full load slip (s_1) as T_1 and the torque at slip $s_2 = 0.3$ as T_2, we have,

$$\frac{T_1}{T_2} = \left(\frac{N_1}{N_2}\right)^2 = \left(\frac{245}{175}\right)^2$$

$$= \frac{3V_1^2 \cdot R_2}{s_1\,\omega_s\left[\left(\frac{R_2}{s_1}\right)^2 + X_2^2\right]} \div \frac{3V_1^2 R_2'}{s_2\,\omega_s\left[\left(\frac{R_2'}{s_2}\right)^2 + X_2^2\right]}$$

$$= \frac{s_2\left[\left(\frac{R_2'}{s_2}\right)^2 + X_2^2\right] R_2}{s_1\left[\left(\frac{R_2}{s_1}\right)^2 + X_2^2\right] R_2'}$$

$$= \frac{0.3\cdot 0.03125\left[\left(\frac{R_2'}{0.3}\right)^2 + 0.2535\right]}{0.02\cdot R_2'\,[(1.5625)^2 + 0.2535]}$$

$$3190\, R_2'^2 - 3235\, R_2' + 72.78 = 0$$

$$R_2' = \frac{3235 \pm 3088.13}{6380}$$

$$= 0.0230 \quad \text{or} \quad 0.99108$$

The first value is less than the original resistance and hence not applicable.

Therefore, the resistance to be added with each slip ring

$$= \frac{0.99108 - 0.03125}{(1.25)^2}$$

$$= 0.6143 \text{ ohm.}$$

(v) *Injection of voltage in rotor circuit*: Let us assume that the induction motor is driving a constant torque load at constant voltage and constant frequency of supply. Neglecting the voltage drop in the stator, the applied voltage per phase $V_1 \simeq E_1 = K\phi =$ constant, irrespective of other operating conditions of the motor. The induced emf in the rotor, when the motor runs with a slip s has a magnitude sE_2, where E_2 is the induced emf in the rotor at standstill conditions of the rotor. The magnitude of the actual rotor current,

$$I_2 = \frac{s\,E_2}{\sqrt{R_2^2 + (s\,X_2)^2}}.$$

Since $R_2^2 \gg (s X_2)^2$, $\bar{I}_2$ phasor will be in phase with $\bar{E}_{2s}$ and $| I_2 | = \frac{s\,|E_2|}{R_2}$. Now, suppose that an additional emf $\bar{E}_j$, opposite in phase to E_{2s}, is injected into the rotor circuit. Initially, when the speed cannot change due to the inertia of the rotor, the net emf in the rotor circuit reduces to a value $(sE_2 - E_j)$, as a result of which the rotor current I_2 and, hence, the torque developed decreases. But, since the load torque remains constant, the speed of the motor starts decreasing. This process of reduction in speed (increase in slip) continues till the rotor induced emf increases to circulate enough current in the rotor to develop the desired torque.

Let s_j be the new value of the slip and $s_j E_2$, the corresponding new value of the rotor emf, once steady state conditions have been reached after the injection of the additional emf E_j. Then,

$$I_2 \simeq \frac{s_j E_2 - E_j}{R_2} \text{ since } (s_j X_2)^2 \ll R_2^2.$$

This must be equal to the original rotor current, since both flux and developed torque are constant. Therefore,

$$\frac{s_j E_2 - E_j}{R_2} = \frac{s E_2}{R_2}$$

or

$$s_j = s + \frac{E_j}{E_2} \tag{4.25}$$

i.e. when the emf injected is in phase opposition to the rotor induced emf, the slip increases or the speed of the motor decreases.

By similar reasoning, it is easy to observe that when the injected emf is in phase with the rotor induced emf, the slip decreases or the speed of the motor increases. Eqn. (4.25), under such conditions, will be expressed as

$$s_j = s - \frac{E_j}{E_2}.$$

If $E_j/E_2 > s$, the new slip s_j becomes negative, i.e. the machine runs at a speed greater than synchronous speed, maintaining its motor operation. The modified speed torque characteristics are shown in Fig. 4.7.

In order to inject the desired emf at slip frequency, a rotating frequency converter is used. Fig. 4.8 shows a simplified sketch of a frequency converter. The stator has no winding and the stator iron structure itself is there only to offer a low reluctance path to the magnetic flux. The rotor consists of a dc armature winding fitted with both sliprings at one end and commutator at the other end.

When three phase currents of frequency f are fed to the sliprings, a rotating magnetic field is produced. The speed of rotation of this field is $N_s = \frac{120f}{P}$ rpm relative to the armature conductors, irrespective of the speed of rotation of the armature. If the armature were stationary, the

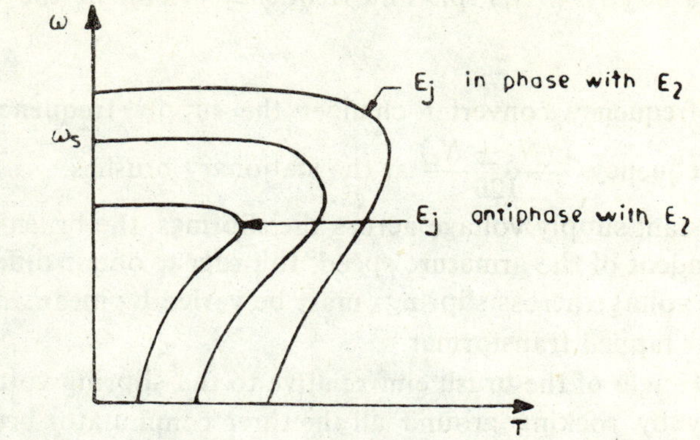

Fig. 4.7. Speed-torque curves with injection of rotor voltage.

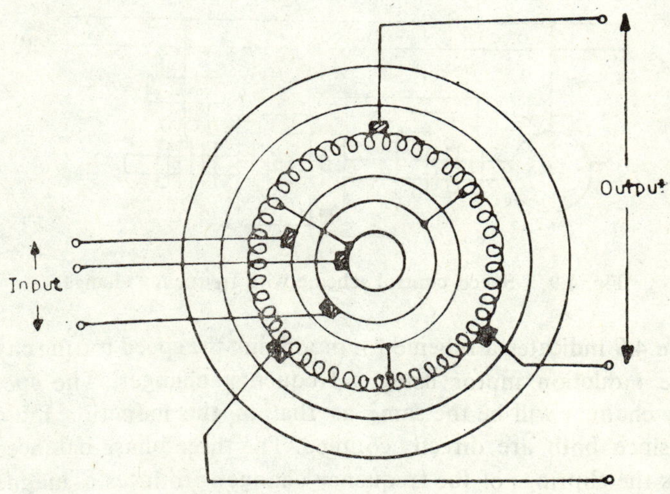

Fig. 4.8 Schematic diagram of frequency converter.

speed of the magnetic field in space will be N_s. Hence, the speed of the field relative to the brushes, which are stationary in space will be N_s itself and the frequency of emf available at the brushes will be given by $\frac{N_s P}{120}$, i.e. f itself.

Now, if the armature were rotated at a speed N_r rpm against the direction of rotation of the rotating magnetic field, the speed of the field in space or the speed relative to the fixed brushes would be $(N_s - N_r)$ rpm, so that the frequency of the emf at the brushes would be $\frac{P(N_s - N_r)}{120}$.

If the armature were to revolve at speed N_r in the same direction as that of the magnetic field, the field speed in space or relative to the stationary brushes would be $(N_s + N_r)$ rpm and frequency of emf at the brushes be $\frac{P(N_s + N_r)}{120}$.

Thus, the frequency converter changes the supply frequency f at the sliprings to frequency $\frac{P(N_s \pm N_r)}{120}$ at the stationary brushes.

For a constant supply voltage across the sliprings, the brush emf is fixed and is independent of the armature speed. In order to obtain different values of brush emf, voltage across sliprings must be varied by means of an auto-transformer or tapped transformer.

The phase angle of the brush emf relative to the slipring voltage can be varied, merely by rocking around all the three commutator brush sets together to a new spatial position.

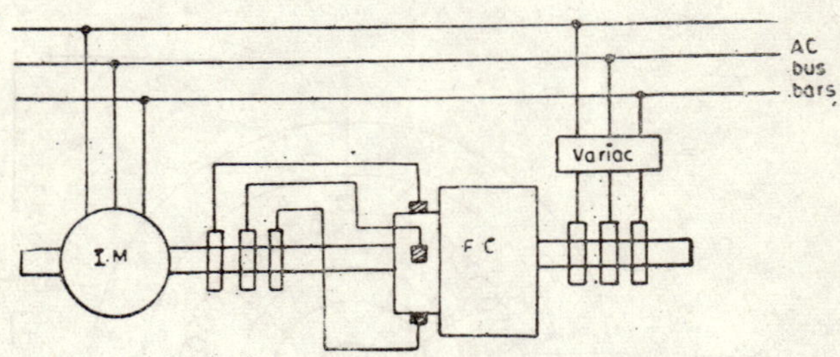

Fig. 4.9. **Speed control scheme with frequency changer.**

Figure 4.9 indicates a scheme for modifying the speed torque characteristic of the induction motor using a frequency changer. The speed of the frequency changer will be the same as that of the induction motor under control, since both are directly coupled. The three phase balanced voltage fed across the sliprings of the frequency changer produces a magnetic field which rotates at speed N_s with respect to the rotor of the frequency changer. If this magnetic field were to rotate in a direction opposite to that of the motor shaft, the speed of the field in space (i.e. with respect to the brushes) would be $(N_s - N_r) = sN_s$. Hence, the frequency of emf available across the brushes pressing over the commutator of the freqnency changer will always be equal to the slip frequency, irrespective of the speed of rotation of the motor.

It can be observed that the slip frequency power is returned to or drawn from the supply through the frequency changer at subsynchronous or supersynchronous speeds respectively. In fact, there are several schemes for recovering the slip frequency power and all of them comprise a means for injecting adjustable voltages of slip frequency into the rotor circuit of a slipring induction motor.

(vi) *Pole changing*: It is well known that a squirrel cage type of rotor winding is one, which is not wound for any specific number of poles and that it adapts to the same number of poles as the airgap magnetic field which is determined by the stator winding. It is possible to change the number of magnetic poles by a factor of 2 for a single stator winding by suitably reconnecting coil groups at the terminals of the winding. In this way, two different synchronous speeds are obtained. This type of winding is called the consequent pole winding and four synchronous speeds may be obtained in a single motor that has two distinct stator windings.

The basic principle of pole changing can be explained as given below: Fig. 4.10(a) shows 6 coils belonging to a particular phase and carrying currents in the directions shown. It is seen that a 12 pole magnetic field is produced by such an arrangement. Coils 1, 3, 5, and 2, 4, 6, are connected in series constituting two distinct coil groups a-b and c-d. If the terminals b and c are connected as in Fig. 4.10(b) the six coils become in series. If the terminal a is connected to c and b to d, coil groups a-b and c-d are connected in parallel (Fig. 4.10(c)). If the direction of currents in coils 2, 4 and 6 (or 1, 3 and 5) were reversed as shown in Fig. 4.11(a), a 6 pole magnetic field would be produced and the synchronous speed would be doubled. Here again, it is possible to connect both the coil groups a-b and c-d in sreies (Fig. 4.11 (b)) or parallel (Fig. 4.11 (c)).

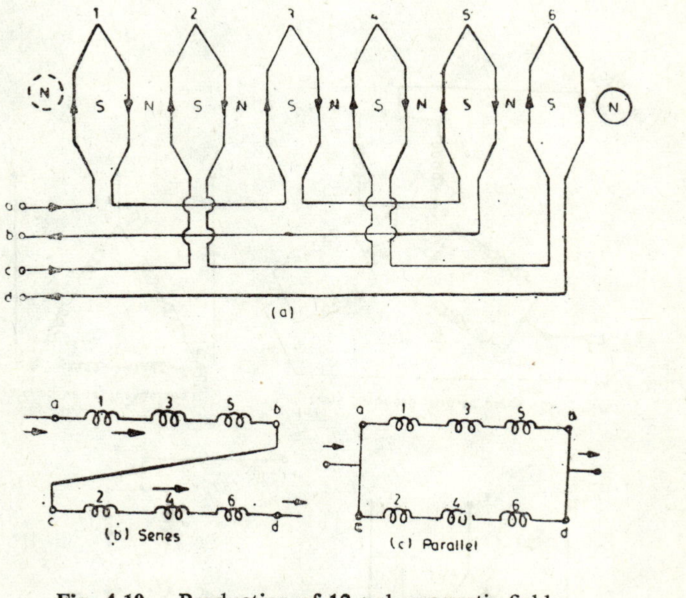

Fig. 4.10. Production of 12-pole magnetic field.

In Figs. 4.10(a) and 4.11(a), six coils belonging to one phase only have been shown. By use of star and delta types of connections for the three phases in combination with series and parallel arrangement of coil groups, different speed torque characteristics can be obtained. Three such possibilities, involving the two speeds obtained from a single consequent, pole type

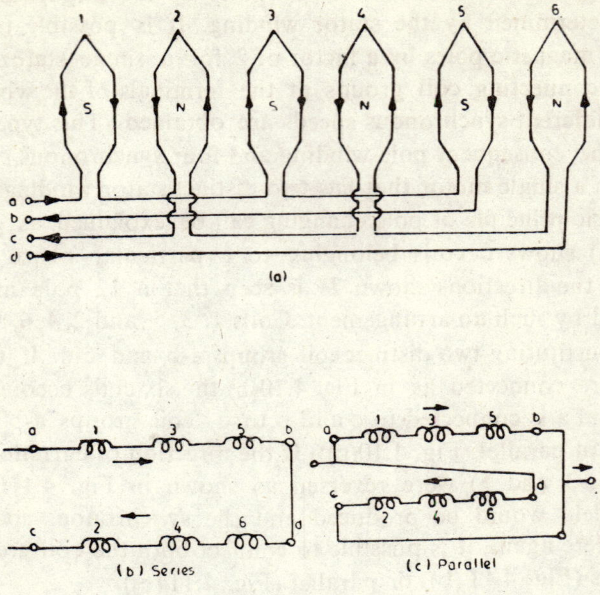

Fig. 4.11. Production of 6-pole magnetic field.

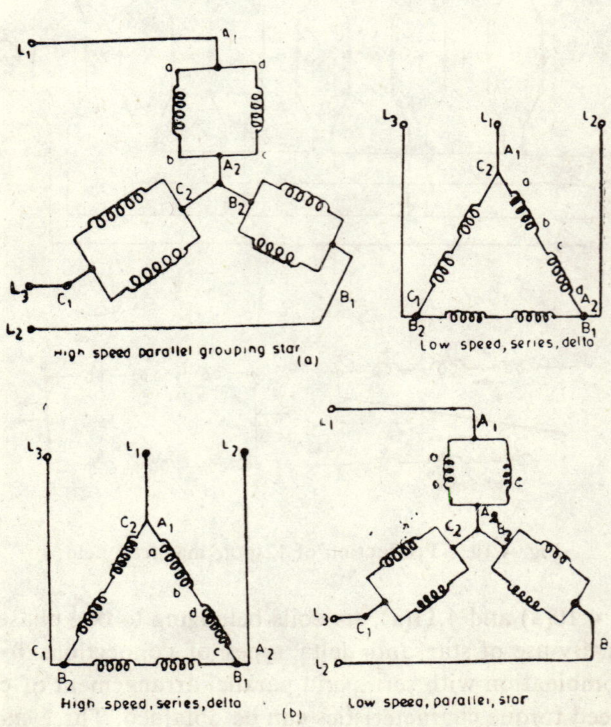

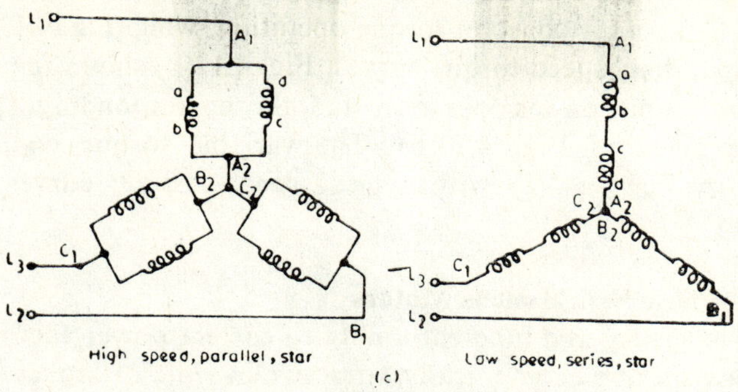

Fig 4.12. Connections for two speed operation with
(a) constant torque
(b) constant power output
(c) variable torque.

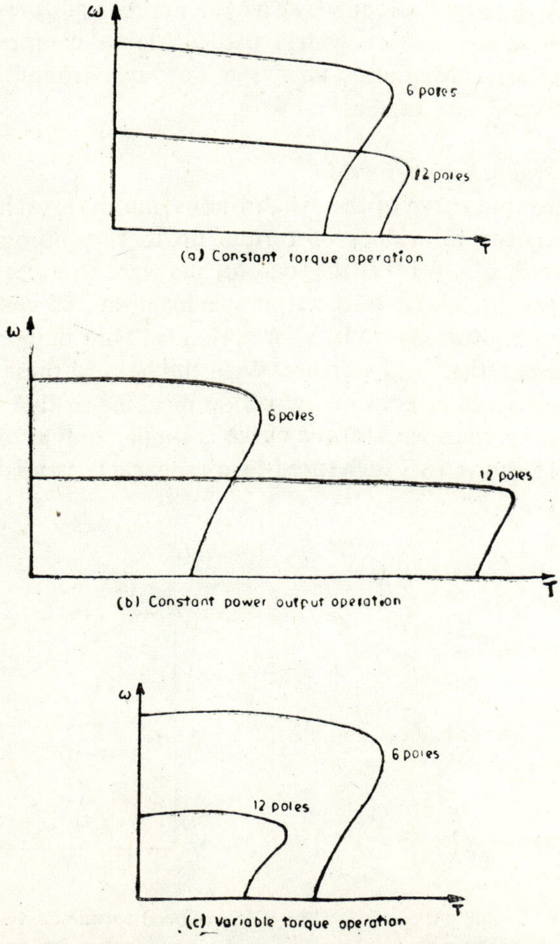

Fig. 4.13. Speed-torque curves of pole changing motors.

of winding, are shown in Fig. 4.12. Fig. 4.12(a) shows the low and high speed connections for constant torque operation, while Fig. 4.13(a) presents the corresponding speed torque curves. Fig. 4.12(b) shows the connections for constant output power operation and the corresponding speed-torque curves are given in Fig. 4.13(b). The variable torque connections are presented in Fig. 4.12(c) with typical speed torque curves shown in Fig. 4.13(c).

4.2 Three-Phase Synchronous Motors

Higher efficiency and inherent ability to correct power factor can make synchronous motors economically attractive in spite of higher capital cost. Also, with size requirements above 20000 kW at 1500 rpm, induction motors are simply not available. Size is not necessarily a limitation for the synchronous motor. Two pole machines have been constructed well above 80000 kW.

A synchronous motor normally runs at a constant speed whose magnitude is fixed by the supply frequency and the number of poles. As constant speed drives, these motors are widely used in large compressors, pumps and stone crushers. Medium and large size synchronous motors find application in paper and cement industry.

4.2.1 Steady state characteristics

The speed-torque curve of the synchronous motor is a horizontal line (constant speed) for all values of torque up to the pull-out (maximum) value, as shown in Fig. 4.14. If the load torque were to exceed the value of maximum torque, the motor falls out of synchronism and comes to a stop.

Synchronous motors, nowadays, are provided with damper windings in the rotor to make them self-starting. With the help of these windings, the motor starts and accelerates as an induction machine so that for all speeds below synchronous the speed-torque curve is similar to that of an induction motor. Fig. 4.15 shows two such speed-torque characteristics during starting

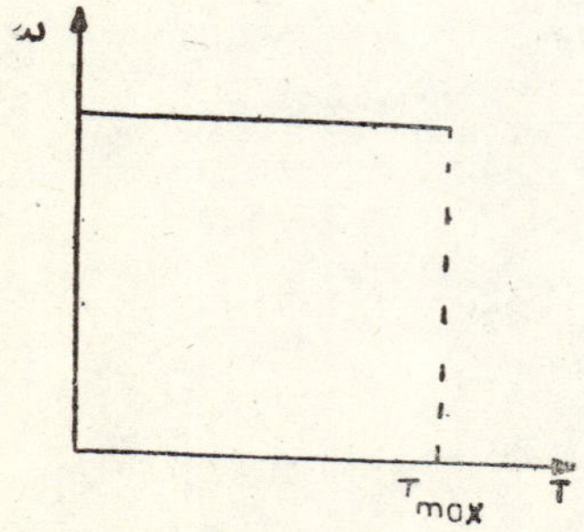

Fig. 4.14. Speed-torque curve of synchronous motor.

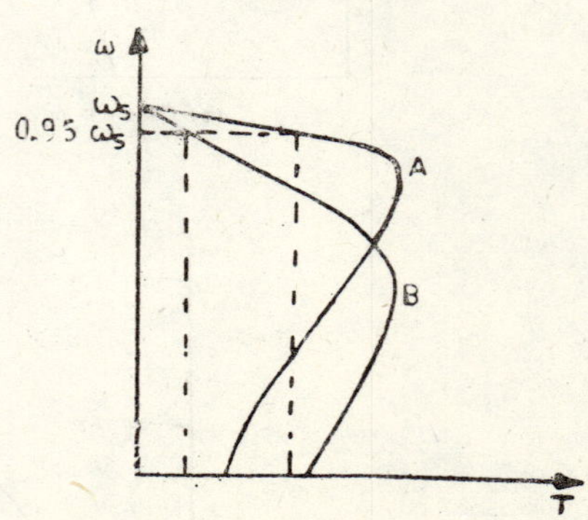

Fig. 4.15. Speed-torque curves of synchronous motor during starting.

of synchronous motor. Curve A corresponds to a motor having less starting torque and considerable pull-in torque (the torque that is developed at nearly 95 per cent of the synchronous speed, when, usually, the dc excitation is switched on).

If the damper winding were to have larger resistance than the previous case, the starting torque would be more, but the pull-in torque will be less, as shown by curve B in Fig. 4.15. The choice of a motor having one or other type of characteristic during starting, simply depends on the type of load to which the motor shaft is connected with.

During steady state, with varying loads, rotor speed oscillations take place over a mean value. These oscillations are due to the variation in the torque angle of the machine. The rotor speed oscillations are of importance while studying the operation of synchronous motor subjected to pulsating loads, for example, as in reciprocating compressors. In order to study even steady state operation of such drives, it is necessary to know the relationship between the electromagnetic torque developed by the motor and the torque angle, i.e. $T = f(\delta)$.

4.2.2 *Torque Angle Characteristic of a Synchronous Motor*

The torque versus torque angle curve of a synchronous motor can be deduced from the simplified phasor diagram of the motor, shown in Fig. 4.16. This diagram shows the direct and quadrature axis components of the motor current and is drawn on the assumption that the stator resistance is negligible. The following equations may be written by inspection of the phasor diagram.

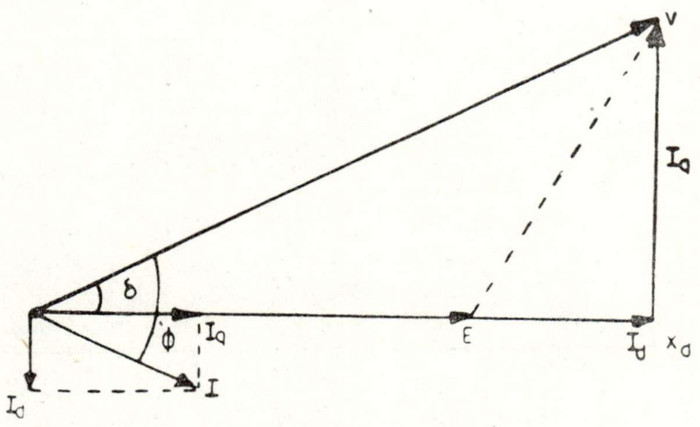

Fig. 4.16. Simplified phasor diagram of salient pole synchronous motor.

$$V \cos \delta = E + I_d X_d \qquad (4.26)$$

$$V \sin \delta = I_q X_q \qquad (4.27)$$

$$I^2 = I_d^2 + I_q^2 \qquad (4.28)$$

Also, the input power is given by

$$P = 3\ VI\cos\phi, \tag{4.29}$$

where P is the total power input,

V is the applied voltage per phase,

E is the induced emf per phase in the stator,

X_d and X_q are the direct and quadrature axis components of synchronous reactance/phase,

$$(X_d = 0.6 - 1.45 \text{ p.u.}; \quad X_q = 0.4 - 1.00 \text{ p.u.})$$

I_d and I_q are the direct and quadrature axis components of the motor current,

ϕ is the power factor angle and

δ is the torque angle.

Eliminating the currents in the above equations, we get

$$P = 3\left[\frac{VE\sin\delta}{X_d} + \frac{V^2\sin 2\delta}{2}\left(\frac{1}{X_q} - \frac{1}{X_d}\right)\right] \text{ watts} \tag{4.30}$$

Hence

$$T = \frac{3}{\omega_s}\left[\frac{VE\sin\delta}{X_d} + \frac{V^2\sin 2\delta}{2}\left(\frac{1}{X_q} - \frac{1}{X_d}\right)\right] \text{ Newton-metres} \tag{4.31}$$

From Eqn. (4.31), it follows that for a salient-pole synchronous motor, besides a torque which varies sinusoidally with torque angle [Fig. 4.17(a)], there exists an additional second harmonic torque [Fig. 4.17(b)].

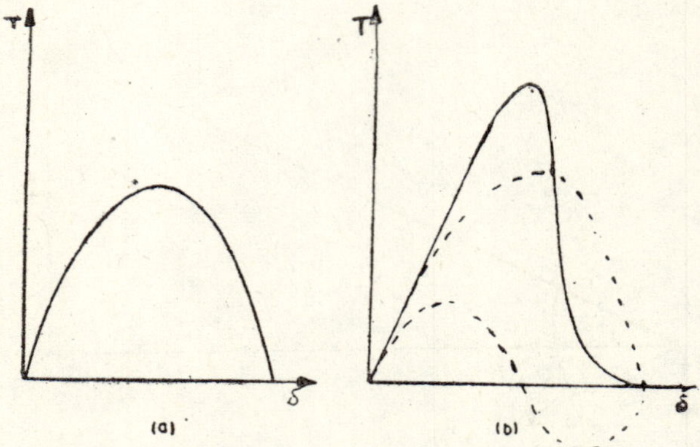

Fig. 4.17. Torque angle curves of synchronous motors.

The first term in the above equation represents the torque due to excitation and the second term is the contribution due to the salient poles. The latter is called reluctance torque, since it is due to the variation in reluctance of the magnetic circuit. Its magnitude will be zero for a cylindrical rotor machine. Thus, the torque developed by the cylindrical rotor synchronous motor is given by

$$T = \frac{3VE \sin \delta}{\omega_s X_d} \tag{4.32}$$

$$= T_{max}.\sin \delta,$$

where T_{max} for a particular motor depends on the induced emf E.

Since with a fixed excitation (E = const.), the torque developed by the motor is directly proportional to the applied voltage, therefore, it is less sensitive to variations in supply voltage than that of an induction motor, whose torque is proportional to the square of the applied voltage.

4.2 3 Modified Speed Torque Characteristics of Three Phase Synchronous Motors

It is well known that the synchronous motor admirably fulfils the requirements of an absolutely constant speed drive, as the speed of the motor depends only on the frequency of the supply voltage and the number of poles for which the motor is wound. The only means to change the synchronous speed of a specific motor is to feed it with a variable frequency supply. In fact, nowadays, the field of application of synchronous motors have been extended by using them together with variable frequency sources, which make the motors adjustable definite speed prime movers. But, it must be borne in mind that, just as in the case of induction motors a variation in the frequency of the source will result in a corresponding change in the flux in the airgap. Hence, in order to operate the motor with fairly constant flux in the airgap, it is necessary to vary the magnitude of the applied voltage in the same ratio as the frequency of the supply (i.e V/f should be kept constant) and to keep the excitation current constant. It can be found from Eqn. (4.31) that this adjustment results in a torque developed by the motor independent of supply frequency, but remaining as a function of torque angle.

Example 4-6. The full load torque angle of a three phase cylindrical rotor synchronous motor at normal voltage and frequency is 45 electrical degrees. If the field current is kept constant, how would the torque angle be affected by the following changes in operating conditions? Ignore the effects of stator resistance, stator leakage reactance and rotational losses of the motor.

(i) Both frequency and supply voltage reduced 5 per cent, load torque on the shaft remaining the same, and

(ii) Both frequency and supply voltage reduced 5 per cent, load power on the shaft remaining the same.

Solution: (i) The torque developed by the motor is given by

$$T = \frac{3VE \sin \delta}{\omega_s X_d}$$

Since E, ω_s and X_d are directly proportional to ω, the angular frequency of the supply and since V is reduced in the same proportion as frequency, for a constant torque to be developed by the motor to cater to the constant load torque, sin δ must remain the same, i.e. δ must remain the same. Hence $\delta = 45°$.

(ii) Power developed by the motor

$$P = \frac{3VE \sin \delta}{X_d}$$

In order to have the same power developed, sin δ must increase $\left(\frac{1}{0.95}\right)$ times, i.e.

$$\text{new } \sin \delta = \sin 45°/0\ 95$$

Therefore, new $\delta = 48.10°$.

PROBLEMS

1. When balanced rated voltage at rated frequency is fed to the stator terminals of a 3-phase squirrel cage induction motor, it develops a maximum torque of 200 per cent of full load torque at a slip of 40 per cent. Neglect stator resistance and rotational losses. Sketch the speed-torque curve for this induction motor.
2. A 10 KW, 440 V, 3-phase, 50 Hz squirrel cage motor operating at rated voltage and frequency has the rotor copper loss at maximum torque 8 times that at full load torque. The slip at full load torque is 0.04. Stator reistance and rotational losses may be neglected.
   Determine:
   (a) the slip at maximum torque,
   (b) the ratio of maximum torque to full load torque, and
   (c) the ratio of starting torque to full load torque.
3. A 3-phase squirrel cage induction motor has a starting current of 6 times its full load current. The motor has a full load slip of 4 per cent. Calculate
   (a) the starting torque,
   (b) the slip at which maximum torque occurs, and
   (c) the ratio of maximum torque to full load torque.
4. A 4-pole, 50 Hz, 3-phase squirrel cage induction motor has a rotor resistance and standstill reactance referred to stator of 0.2 ohm and 0.8 ohm per phase respectively. Its full load slip is 4 per cent. Neglect stator resistance and leakage reactance. How much stator voltage should be reduced to get a speed of 1200 rpm if
   (a) load torque remains constant, and
   (b) load torque varies as the square of the speed.
5. A 6-pole, 50 Hz squirrel cage induction motor has rotor resistance and standstill reactance referred to stator of 0.2 ohm and 1 ohm per phase, respectively. With rated voltage and rated frequency it runs at full load with 4 per cent slip. Neglect stator resistance and rotational losses. Determine the operating speed of the motor, when the stator voltage impressed is reduced to $\frac{1}{\sqrt{2}}$ times the rated voltage, frequency remaining the same, if
   (a) the load torque remains constant at the rated motor torque, and
   (b) the load is of fan-type having rated torque at rated speed.

6. A 3-phase stator connected, 400V, 50 Hz, 6-pole induction motor has the following equivalent circuit constants in ohms per phase.

$$R_1 = 0.6;\ R_2 = 0.2;\ X_1 = X_2 = 1;\ X_m = 100.$$

If the motor were operated at 200 V, 25 Hz, calculate

(a) the current and power factor at the instant of starting and at speed corresponding to maximum torque,
(b) the starting and maximum torques.

Compare your answers for (a) and (b) with those corresponding to rated voltage, rated frequency operation.
Under normal operation, the motor has a rated slip of 0.05. Determine the slip corresponding to the reduced voltage and reduced frequency operation, the load torque remaining the same as its original rated value.

7. Repeat problem 6, when the motor operates at 40 V, 5 Hz instead of 200 V, 25 Hz.

8. Assuming that sufficient balanced impedances are added in series with the stator windings to produce some given reduction in starting torque of a three-phase induction motor, show that for all speeds between zero and synchronous the torque developed when inductances are used exceeds that developed when resistances are inserted.
Will there be any change in the value of the slip corresponding to maximum torque

(a) when inductances are added, and
(b) when resistances are added.

9. A 4-pole, 50 Hz, slip ring induction motor has rotor resistance and standstill reactance referred to stator of 0.2 ohm and 1 ohm per phase respectively. At full load it runs at 1440 rpm. Determine the value of resistance to be inserted in rotor in ohms per phase to operate at a speed of 1200 rpm, if

(a) the load torque remains constant, and
(b) the load torque varies as the square of the speed.

Neglect stator resistance and leakage reactance. Assume a ratio of stator to rotor turns of 1.1.

10. A 10 kW, 400 V, 3-phase, 6-pole, 50 Hz slipring induction motor has a maximum torque equal to twice the full load torque at a slip of 20 per cent, with sliprings short circuited. Neglecting stator resistance and rotational losses, determine the slip at full load, starting torque and torque developed at a slip of 0.25.
If the rotor circuit resistance is increased to four times by connecting resistances in series with each rotor slipring, determine,

(a) the slip at maximum torque,
(b) the slip at which the motor will develop the same full load torque, and
(c) the starting torque

11. Repeat the latter part of the problem 10 for a rotor circuit resistance equal to eight times the inherent rotor resistance. Compare the answers and comment on the results obtained.

12. A 50 kW, 3 phase, 400 V, 50 Hz slipring induction motor has a rotor impedance of $(0.05 + j\,0.4)$ ohm per phase at 50 Hz. The standstill rotor emf is 100 V and the rotor current is kept at 100 A per phase. Determine the speed of the motor, when

(a) the rotor is short circuited,
(b) 20 V per phase is injected in phase opposition to rotor emf, and
(c) 20 V per phase is injected in phase with rotor emf.

13. A 3-phase salient pole synchronous motor has the following per unit values for

its reactances: $X_d = 1.2$ and $X_q = 1.0$. Calculate the emf induced due to excitation when the motor draws rated current at leading power factor and at rated voltage, supplying a mechanical load corresponding to 0.9 per unit power. Neglect all losses.

14. A 3-phase, 400 V, star connected synchronous motor with $X_d = 5$ ohms/phase and $X_q = 4$ ohms/phase is connected directly to the supply network. Calculate the maximum load power that could be put on the shaft, if the field current of the motor were reduced to zero.

15. The synchronous motor of problem 14 drives a load having a constant power of 30 kW. Determine the maximum stable values of torque angle for the given load power and applied voltage at minimum value of the emf induced due to excitation. Calculate, also, the minimum stator induced emf due to field current corresponding to the above load.

Answers

2. (a) 0.155 (b) 2.0656 (c) 0.625
3. (a) 1.25 T_{fl} (b) 0.253 (c) 2.628
4. (a) 0.568 V (b) 0.4734 V
5. (a) 906 rpm (b) 918 rpm
6. (a) 89.7 A, 0.6182; 56.69 A, 0.8679
   (b) 92.2Nm, 214 Nm
   At rated voltage and rated frequency
   (a) 106.7 A, 0.3677; 68.67 A, 0.8
   (b) 65.2 Nm, 280 Nm
   No steady operation is possible at reduced voltage and frequency, since T_{max} is less than T_{rated}.
7. (a) $I_{st} = 27.83$ A, p.f. $= 0.9585$
   $I_{maxT} = 18.42$ A, p.f. $= 0.982$
   (b) $I_{st} = 44.38$ Nm, $T_{max} = 61.33$ Nm
   No operation possible.
9. (a) 0.8 ohm (b) 1.27 ohms
10. $s_{fl} = 0.053$, $T_{st} = 77.7$ Nm T = 197 Nm
    (a) 0.8 (b) 0.2143 (c) 197.03 Nm
11. (a) 1.6 (b) 0.428 (c) 181.47 Nm.
12. (a) 945 rpm (b) 652.2 rpm (c) 1128.3 rpm
13. 2.4 p.u. (phase)
14. 4 kW
15. 76° and 362 V (line)

Chapter 5

Starting

The most important processes associated with a controlled electrical drive are: (i) starting, (ii) speed control or variation of speed of rotation, (iii) braking and (iv) reversing the direction of rotation. As a rule, all these are transient processes that occur in a drive system.

Starting of an electrical drive involves a change in its state from rest to a steady state speed of rotation. The process of starting is the most important phenomenon in the entire operation of the drive. Control of the starting process essentially consists of controlling the acceleration of the driving motor and the latter is basically a problem of modifying the speed torque characteristics of the motor in such a way as to obtain the desired starting performance.

5.1 Effect of Starting on Power Supply, Motor and Load

While studying starting of electric drive systems, it is necessary to consider three factors: (i) effect of starting upon the power supply, (ii) effect of starting upon the driving motor itself, and (iii) effect of starting upon the connected mechanical load.

The supply network to which the motor is connected may affect the selection of the starting device from the following viewpoint. The excessive voltage drop due to the peak starting current may interfere with the supply in such a way that it cannot be tolerated by other equipment or other consumers connected to the same power supply network.

Since starting is associated with excessive currents, the effect of starting upon the motor itself must be carefully considered. The starting currents will add to the motor heating by an amount that depends upon their rms values and upon the frequency of starting. In a dc motor the limitation may be good commutation rather than heating, as dc machines have a certain maximum limit for the current dictated by the commutation process.

The equipment connected to the driving motor may impose strict constraints upon the type of accelerating cycle and upon the maximum permissible acceleration. Discomfort to passengers in lifts and trains may very well set an upper limit to the rate of acceleration. Cranes, excavators and similar material handling equipment must be so equipped that the operator has precise control over the load, whether it be light or heavy. In case of loads, which exhibit stiction, the motor must develop sufficient torque to start the rotation, after which the torque must perhaps be reduced for limiting the acceleration to a safe value.

5.2 Methods of Starting Electric Motors

The different methods of starting of the various types of electric motors would have been dealt with in detail in a course on 'Rotating electrical machines' and, hence, they have been just listed here for the sake of continuity of thought.

1. *Full voltage starting*: This involves the application of full line voltage to the motor terminals. This is also called 'direct-on-line-starting'. D.C. motors upto 2 kW and squirrel cage induction motors as well as certain small synchronous motors upto 4 or 5 kW are usually line-started.

2. *Reduced voltage starting*: In order to avoid heavy starting current and the consequent voltage dip in the supply lines majority of motors are started by applying a reduced voltage to their terminals and subsequently increasing it to its normal value.

The starting of a dc motor is, often, accomplished by the addition of suitable external resistance in the armature circuit and the starting controller is arranged so that this resistance is short-circuited in steps as the motor comes upto speed.

Reduced voltage starting of induction motors is achieved by (i) stator resistance starting, (ii) stator reactor starting, (iii) star-delta starting, and (iv) autotransformer starting.

The above methods apply equally well to the synchronous motors. With reduced voltage starting the transition to full voltage may be made either before or after synchronization although the former method is usually preferred.

It may be noted that the torque at starting, in all the above methods, gets reduced.

3. *Increased torque starting*: With a wound rotor induction motor, resistance can be added in the rotor circuit so as to decrease the starting current while increasing the starting torque, even, upto the value of maximum torque that can be developed by the motor.

4. *Starting by means of smooth variation of voltage or frequency*: With ac motor-dc generator sets, dc motors can be started by smooth variation of applied voltage and with variable frequency sources both induction and synchronous motors can be started by smooth variation of supply frequency, simultaneously varying proportionally the applied voltage to the motors.

Example 5-1. A 3-phase, 100 kW, 400 V, 6-pole, 50 Hz induction motor runs at 950 rpm on full load. If it takes 1203 A on direct on line starting, determine the ratio of starting torque to full load torque, when the motor is:

(i) started direct on line,
(ii) started by using a star delta starter,
(iii) started by an autotransformer with 70 per cent tapping, and
(iv) started by stator resistance, limiting the starting current to 401 A.

Assume that the full load efficiency and power factor of the motor are 0.9 and 0.8 respectively.

Solution:

$$\text{Full load current drawn by the motor} = \frac{100 \cdot 1000}{\sqrt{3} \cdot 400 \cdot 0 \cdot 9 \cdot 0 \cdot 8}\text{ A}$$
$$= 200.5 \text{ A.}$$

$$\text{Full load slip} = \frac{1000-950}{1000}$$
$$= 0.05.$$

From Eqn. (4.15), we have $\frac{T_{st}}{T_f} = \left(\frac{I_{st}}{I_f}\right)^2 \cdot s_f$

But, I_{st} is $\propto$ voltage applied to motor and therefore $I_{st} = x \cdot I_{d.o.L}$, where $I_{d.o.L}$ will be the current drawn if the motor were switched on directly to line and x is the fraction of voltage applied to the motor by different methods of reduced voltage starting.

Hence,

$$\left(\frac{T_{st}}{T_f}\right) = x^2 \left(\frac{I_{d.o.L}}{I_f}\right)^2 ; s_t$$

(i) $x = 1; \frac{I_{d.o.L}}{I_f} = \frac{1203}{200.5} = 6$

Hence $\left(\frac{T_{st}}{T_f}\right) = (6)^2 \cdot 0.05 = 1.8$

(ii) When started by means of star-delta starter, the voltage applied will be across star connected windings and hence $x = \frac{1}{\sqrt{3}}$.

Therefore, $\frac{T_{st}}{T_f} = \left(\frac{1}{\sqrt{3}}\right)^2 \cdot 1.8 = 0.6.$

(iii) $x = 0.7.$

Hence, $\frac{T_{st}}{T_f} = (0.7)^2 \cdot 1.8 = 0.882$

(iv) $x = \frac{401}{1203} = \frac{1}{3}$

Therefore, $\frac{T_{st}}{T_f} = \left(\frac{1}{3}\right)^2 \cdot 1.8 = 0.2.$

5.3 Acceleration Time

The time required to attain a change in speed from ω_1 to ω_2 may be determined from the equation of motion.

$$T_M = T_L + J \cdot \frac{d\omega}{dt}, \text{ where both } T_M \text{ and } T_L \text{ are functions of } \omega.$$

i.e. $$T_M - T_L = J \cdot \frac{d\omega}{dt}$$

i.e. $$dt = J \cdot \frac{d\omega}{T_M - T_L}$$

Therefore, $$t = J \int_{\omega_1}^{\omega_2} \frac{d\omega}{T_M - T_L} \qquad (5.1)$$

In the above equation the inertia J is assumed to be constant. In case J varies with time or speed, the equation may be rewritten suitably.

For the case, where ω_1 is zero at t equal to zero and ω_2 is equal to a steady state speed ω_0 attained by the drive, the accelerating time is given by

$$t = J \int_0^{\omega_0} \frac{1}{T_M - T_L} \cdot d\omega. \qquad (5.2)$$

The value of the integral in Eqn. 5.2 is the area under the curve relating $\left(\frac{1}{T_M - T_L}\right)$ to ω. For a typical motor load combination, whose speed torque characteristics are shown in Fig. 5.1(a), the reciprocal of the accelerating torque $1/(T_M - T_L)$ plotted as a function of speed is given in Fig. 5.1(b). The shaded area is the value of the integral in question and when multiplied by total inertia J gives the accelerating time. It should be noted that the reciprocal of $(T_M - T_L)$ would become infinitely large as the

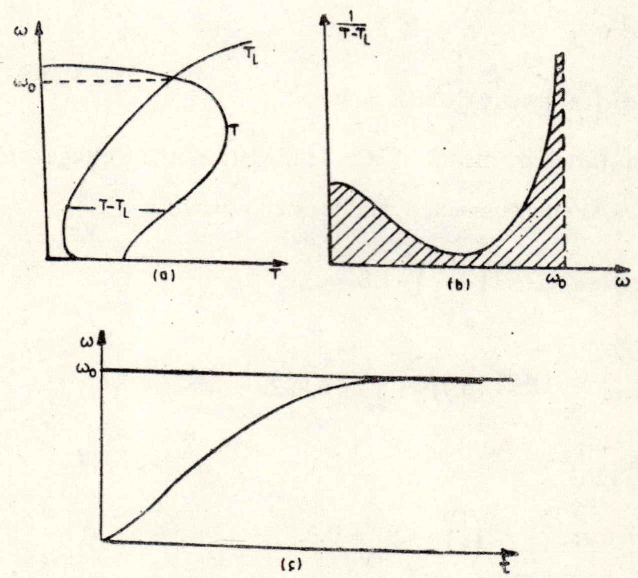

Fig. 5.1. (a) Speed-torque curves of motor and load
(b) Area under the curve $1/T_M - T_L$ vs ω
(c) Speed time curve.

speed approaches its steady state value, and this would lead to an infinite accelerating time. This difficulty is removed, in practice, by computing the time required to reach say 95 to 98 per cent of the final steady state speed. The desired value of accelerating time for any motor load combination can be achieved by suitably modifying the speed torque characteristic of the driving motor by means of any one of the methods studied earlier.

It should also be observed that the speed versus time curve can be constructed from the relationship between the reciprocal of the accelerating torque and speed. The time required to reach any speed ω_x is given by the expression

$$t_x = J\int_0^{\omega_x} \frac{1}{(T_M - T_L)}.d\omega \tag{5.3}$$

The resulting motor speed-time curve for the drive system characterised in Fig. 5.1(a) is presented in Fig. 5.1(c).

Example 5-2. A 3-phase, 440 V, 50 Hz induction motor driving a fan, whose torque is proportional to the square of the speed, has the speed-torque characteristic given in the following table:

Speed (rpm)	1500	1425	1350	1200	1050	900	750	600	450	300	150	0
Torque (N-m)	0	42	60	70	68	60	50	40	33	29	28	27

Determine the time taken by the fan having a rated torque of 42 N-m to accelerate to its rated speed of 1425 rpm. The total moment of inertia of the motor and fan may be taken as 1 kg-m².

Solution: It is possible to straight away plot $\left(\frac{1}{T_M - T_L}\right)$ versus speed curve for the drive system under consideration using the following table:

Speed N rpm	Motor torque T N-m	Fan torque $T_L = \frac{T_L \text{ rated}}{N_{(\text{rated})}^2} \times N^2$	$\frac{1}{T_M - T_L}$
1425	42	42	∞
1350	60	37.7	0.045
1200	70	29.8	0.025
1050	68	22.8	0.022
900	60	16.7	0.023
750	50	11.6	0.026
600	40	7.4	0.031
450	33	4.2	0.035
300	29	1.8	0.037
150	28	0.47	0.036
0	27	0	0.037

Figure 5.2 shows the plot. The area enclosed by the curve upto a speed of 1400 rpm, i.e. about 98.25 per cent of the final speed of 1425 rpm attained by the fan is found to be nearly 155 sq. cm.

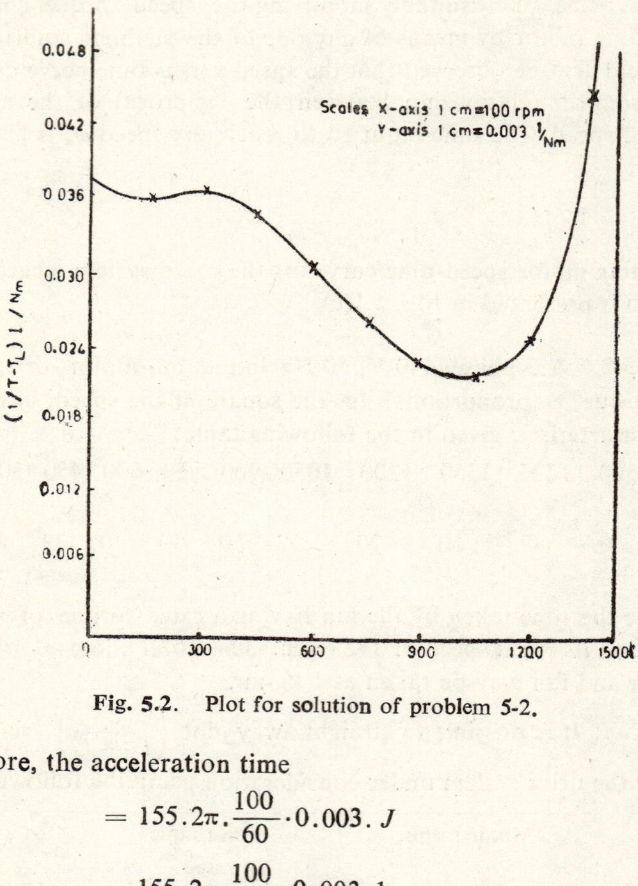

Fig. 5.2. Plot for solution of problem 5-2.

Therefore, the acceleration time

$$= 155.2\pi.\frac{100}{60}\cdot 0.003.\,J$$

$$= 155.2\pi.\frac{100}{60}\cdot 0.003.1$$

$$= 4.87 \text{ secs.}$$

5.3.1 *Acceleration Time for Specific Nature of Motor and Load Torques*

(i) *Constant motor and load torques*: Such a situation may arise, for example, during starting of a hoisting mechanism with a constant load torque. If the motor also develops a constant magnitude of torque during starting (it is possible to have such a special speed torque characteristic with shunted armature connection of a dc series motor or a double cage motor or rotor resistance starting of a slipring induction motor), the net torque available for acceleration $(T_M - T_L)$ will also be a constant. Since $(T_M - T_L)$ is equal to $J\frac{d\omega}{dt}$, $\frac{d\omega}{dt}$ itself will be a constant in this case, i.e. speed rises linearly with time and is given by the expression

$$\omega = \frac{T_M - T_L}{J} . t \tag{5.4}$$

The time t_1 taken by the motor to reach a speed of ω_1 rad/sec is given by

$$t_1 = \frac{J\,\omega_1}{T_M - T_L} \tag{5.5}$$

During starting, let us assume that $T_L = 0$, $T_M =$ rated torque T_f and the moment of inertia J is due only to the rotor of the motor. The motor reaches a speed ω_0 which is the ideal no load speed of the motor. Then,

$$t_1 = \frac{J\,\omega_0}{T_f} \tag{5.6}$$

The time t_1 given by the above expression is called the mechanical time constant of the motor and denoted by t_m. In case of dc motors and induction motors of 1000 rpm speed t_m varies from 0.4—0.6 second.

(ii) *Linearly varying motor torque and constant load torque*: As explained earlier, a linearly varying motor torque, during starting, can be exhibited by a dc series motor with shunted armature or shunt motor connection or a slipring induction motor whose rotor circuit resistance is quite high. At $\omega = 0$, the motor torque corresponds to the starting torque T_{st} and at $\omega = \omega_s$, the synchronous speed of an induction motor $T_M = 0$ (Fig. 5.3). The relationship between motor torque and slip is expressed as:

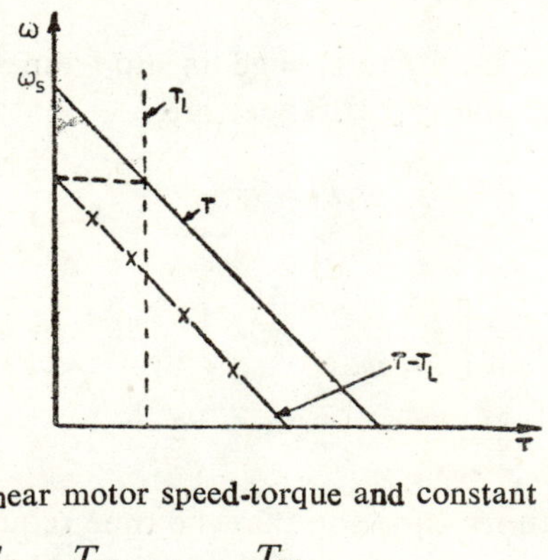

Fig. 5.3. Linear motor speed-torque and constant load curves.

$$\frac{T_{st}}{1} = \frac{T_M}{s} \text{ or } s = \frac{T_M}{T_{st}} \tag{5.7}$$

$$\text{The accelerating torque } J\frac{d\omega}{dt} = T_M - T_L \tag{5.8}$$

Using Eqns. (5.7) and (5.8), we have

$$-\omega_s \frac{ds}{dt} = \frac{sT_{st} - T_L}{J},$$

since $\omega = \omega_s(1-s)$ and hence $d\omega = -\omega_s . ds$.

$$\text{Therefore,} \quad \frac{ds}{dt} + \frac{T_{st}}{J\omega_s} s - \frac{T_L}{J\omega_s} = 0 \tag{5.9}$$

The quantity $\frac{J\omega_s}{T_{st}}$ must have a dimension of seconds and is termed as the starting time constant, t_{st} of the drive. Here, J represents the total moment of inertia of the drive system, including the inertia of any external rotating masses other than the rotor of the driving motor.

Eqn. (5.9), then, becomes

$$\frac{ds}{dt}+\frac{1}{t_{st}}.s-\frac{1}{t_{st}}.\frac{T_L}{T_{st}}=0 \tag{5.10}$$

5.3.2 Optimum Value of s_{maxT} for Minimum Accelerating Time

If load and friction torques were negligible, the equation of motion could be written as

$$T_M = J\frac{d\omega}{dt}$$

$$= J.\frac{d}{dt}[\omega_s(1-s)]$$

$$= -\omega_s J.\frac{ds}{dt} \tag{5.11}$$

Neglecting stator resistance, the electromagnetic torque developed by the induction motor is given by

$$T = T_{max}.\frac{2}{\dfrac{s}{s_{maxT}}+\dfrac{s_{maxT}}{s}} \tag{5.12}$$

The time taken by the motor to change its slip from s_1 to s_2 is obtained by equating Eqns. (5.11) and (5.12). Therefore,

$$-J\omega_s.\frac{ds}{dt} = T_{max}\frac{2}{\dfrac{s}{s_{maxT}}+\dfrac{s_{maxT}}{s}}$$

i.e.
$$\int_0^t dt = \frac{-J\omega_s}{2T_{max}}\left[\int_{s_1}^{s_2}\frac{s}{s_{maxT}}ds+\int_{s_1}^{s_2}\frac{s_{maxT}}{s}.ds\right]$$

i.e.
$$t = \frac{J\omega_s}{2T_{max}}\left[\frac{s_1^2-s_2^2}{2s_{maxT}}+s_{maxT}\log_e\frac{s_1}{s_2}\right] \tag{5.13}$$

It is clear from the above equation that the time required for the slip to change from s_1 to s_2 is a function of the slip at which the maximum torque of the motor occurs. The value of s_{maxT} that would give minimum time to change the slip of the motor from s_1 to s_2 can be obtained by differentiating Eqn. (5.13) w.r.t. s_{maxT} and equating the result to zero.

$$(s_{maxT})_{opt} = \left[\frac{(s_1^2-s_2^2)}{2\log_e(s_1/s_2)}\right]^{1/2} \tag{5.14}$$

The minimum time can be determined by substituting Eqn. (5.14) in Eqn. (5.13).

The optimum value of s_{maxT} to accelerate the motor from $s_1 = 1$ (standstill conditions) to any arbitrary slip s in minimum time is given by

$$(s_{maxT})_{opt} = \left[\frac{1-s^2}{2\log_e(1/s)}\right]^{1/2} \tag{5.15}$$

The optimum value of rotor resistance to change the slip of the motor from s_1 to s_2 in minimum time can be obtained from Eqn. (5.14) and the relation

$$\frac{R_2}{s_{\max T}} = (X_1 + X_2).$$

Hence,

$$[R_2]_{\text{opt}} = (X_1 + X_2)\,(s_{\max T})_{\text{opt}}$$

$$= (X_1 + X_2)\left[\frac{(s_1^2 - s_2^2)}{2\log_e (s_1/s_2)}\right] \qquad (5.16)$$

Example 5-3. A 3-phase, 400 V, 6-pole, 50 Hz, star connected wound rotor induction motor has a sum of stator and rotor leakage reactance referred to stator of 1 ohm. It is connected to a balanced 400 V supply and drives a pure inertia, load. The moment of inertia of rotor including the load is 10 kg-m². Direct on line starting is used and the rotor circuit resistance is adjusted so that the motor brings its load from rest to 0.95 of synchronous speed in shortest possible time.

Neglecting losses except those of the rotor and the exciting current, calculate the value of the rotor resistance referred to the stator and the minimum time to reach 0.95 of synchronous speed.

Solution: The optimum value of $s_{\max T}$ to accelerate the motor from rest to a slip of 0.05 in shortest possible time is given by

$$(s_{\max T})_{\text{opt}} = \left[\frac{1 - s^2}{2\log_e (1/s)}\right]^{1/2}$$

$$= \left[\frac{1 - (0.05)^2}{2\log_e (1/0.05)}\right]^{1/2}$$

$$= 0.408$$

$$(R_2)_{\text{opt}} = (X_1 + X_2).(s_{\max T})_{\text{opt}}$$

$$= 1 \cdot 0.408 = 0.408 \text{ ohm}$$

The minimum time

$$t_{\min} = \frac{J \cdot \omega_s}{2\,T_{\max}}\left[\frac{1 - s^2}{2(s_{\max T})_{\text{opt}}} + (s_{\max T})_{\text{opt}} \log_e \frac{1}{s}\right]$$

$$T_{\max} = \frac{3V_1^2}{\omega_s \cdot 2(X_1 + X_2)}$$

$$= \frac{3 \cdot (400/\sqrt{3})^2}{(2\pi \cdot 1000/60) \cdot 2 \cdot 1}$$

$$= \frac{(400)^2 \cdot 60}{4\pi \cdot 1000} = 763.94 \text{ N-m}$$

Hence,

$$t_{\min} = \frac{10 \cdot (2\pi \cdot 1000/60)}{2 \cdot 763.94}\left[\frac{1 - (0.05)^2}{2 \cdot 0.408} + (0.408)\log_e\left(\frac{1}{0.05}\right)\right]$$

$$= 0.685\,[1.22 + 1.22]$$

$$= 1.67 \text{ secs.}$$

5.4 Energy Relations During Starting

Certain useful expressions regarding the energy losses in a motor during periods of changing speed like starting, braking and sudden changes in applied voltage, load etc. can be obtained as described below:

5.4.1 DC Shunt Motor

An expression of the form given below may be obtained for the energy lost in the armature of a dc shunt motor.

The voltage equation of the dc motor is

$$V = E + I_a R_a \text{ or } I_a R_a = V - E$$

For the shunt motor, $$I_a R_a = V - K\omega \tag{5.17}$$

The above Eqn. (5.17) is valid for a separately excited motor also. The equation of motion on no load will be

$$T_M = K I_a = J \frac{d\omega}{dt} \tag{5.18}$$

Multiplying Eqn. (5.17) by Eqn. (5.18), we get

$$I_a^2 R_a = \frac{J \cdot V}{K} \cdot \frac{d\omega}{dt} - J\omega \frac{d\omega}{dt} \tag{5.19}$$

On no load, the $I_a R_a$ in the motor will be negligibly small and hence $V = K\omega_0$, where ω_0 is the no load speed of the shunt motor.

Substituting $\frac{V}{K} = \omega_0$ in Eqn. (5.19), we get

$$I_a^2 R_a \cdot dt = J\omega_0 \cdot d\omega - J\omega \cdot d\omega \tag{5.20}$$

If the motor speed were to change from ω_1 to ω_2 from time t_1 to t_2, energy dissipated in the armature circuit is given by

$$W = \int_{t_1}^{t_2} I_a^2 R_a \, dt$$

$$= J\omega_0 \int_{\omega 1}^{\omega 2} d\omega - J \int_{\omega 1}^{\omega 2} \omega \cdot d\omega$$

$$= J\omega_0 (\omega_2 - \omega_1) - \frac{J}{2} (\omega_2^2 - \omega_1^2)$$

Hence, energy lost during starting, when the motor changes its speed from zero to no-load speed ω_0 will be,

$$W_{st} = J\omega_2^2 - \frac{J}{2} \omega_2^2 = \frac{J\omega_0^2}{2} \text{ joules} \tag{5.21}$$

Thus, energy lost in the armature of a dc motor (shunt excited or separately excited) during starting on no-load will be equal to the kinetic energy absorbed by the armature, while accelerating from standstill to no-load speed. Further, this energy loss is independent of the armature circuit resistance.

If the motor were started with a constant load torque T_L, the energy lost during starting could be determined as follows:

The equation of motion will, now, be of the form

$$T_M = KI_a = T_L + J \frac{d\omega}{dt} \tag{5.22}$$

Multiplying Eqn. (5.17) by Eqn. (5.22), we get

$$I_a^2 R_a = J \frac{V}{K} \cdot \frac{d\omega}{dt} - J \cdot \omega \cdot \frac{d\omega}{dt} + \frac{V}{K} T_L - T_L \cdot \omega \tag{5.23}$$

If the motor speed were to change from ω_1 to ω_2, the energy lost would be given by,

$$W = \int_{t_1}^{t_2} I_a^2 R_a \, dt$$

$$= J \cdot \frac{V}{K} \int_{\omega_1}^{\omega_2} d\omega - J \int_{\omega_1}^{\omega_2} \omega \cdot d\omega + \frac{V}{K} T_L \int_{t_1}^{t_2} dt - T_L \int_{t_1}^{t_2} \omega(t) \, dt$$

$$= J \cdot \frac{V}{K} (\omega_2 - \omega_1) - \frac{J}{2} (\omega_2^2 - \omega_1^2) + \frac{V}{K} T_L (t_2 - t_1) - T_L \int_{t_1}^{t_2} \omega(t) \, \omega \, dt \tag{5.24}$$

V/K in the above Eqn. (5.24) can be replaced by ω_0, the no-load speed. Therefore, the energy lost during starting on load, when the speed changes from zero to ω_r, will be

$$W_{st} = J\omega_0 \omega_r - \frac{J}{2} \omega_r^2 + T_L \cdot \omega_0 \, t_{st} - T_L \int_0^t \omega(t) \, dt,$$

$$= J\left(\omega_0 \omega_r - \frac{\omega_r^2}{2}\right) + T_L \left(\omega_0 \, t_{st} - \int_0^t \omega(t) \, dt\right) \tag{5.25}$$

where t_{st} is the accelerating time.

The second term in Eqn. (5.25) shows that the energy lost during starting depends both on the accelerating time and on the nature of speed variation with time during acceleration. In fact, the expression

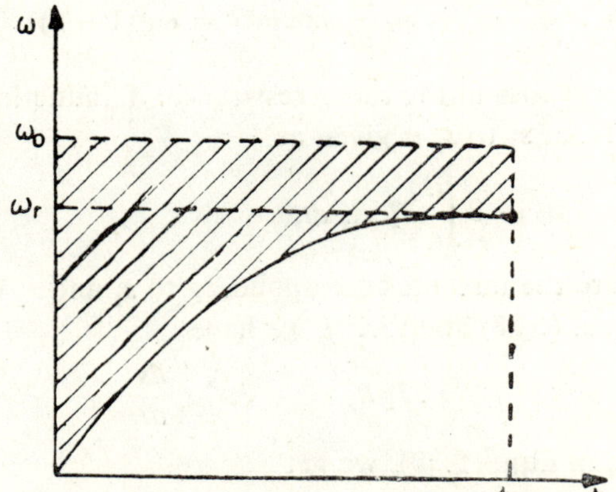

Fig. 5.4. Speed-time curve during starting of a dc shunt motor.

$\left(\omega_0\, t_{st} - \int_0^t \omega(t)\, dt\right)$ can be determined as the shaded area shown in Fig. 5.4.

5.4.2 DC Series Motor

Since, a dc series motor should not be started on no-load, let us assume a constant load torque T_L to be present, while determining the energy loss during starting.

The equation of motion will be,

$$T_M = KI_a^2 = J\frac{d\omega}{dt} + T_L \tag{5.26}$$

Hence, the total energy dissipated in the armature circuit resistance is given by the following expression,

$$W = \int_{t_1}^{t_2} I_a^2 R_a\, dt = \frac{JR_a}{K}\int_{\omega_1}^{\omega_2} d\omega + \frac{T_L}{K} R_a \int_{t_1}^{t_2} dt$$

$$= \frac{J\, R_a}{K}(\omega_2 - \omega_1) + \frac{T_L R_a}{K}(t_2 - t_1) \tag{5.27}$$

Thus, the energy dissipated, while the motor speed changes from ω_1 to ω_2, is proportional to R_a, the armature circuit resistance, rather than independent of this quantity as in the case of a dc shunt motor.

5.4.3 Three Phase Induction Motor

Let us consider, now, a three phase induction motor. The torque developed is given by the expression

$$T = \frac{3}{\omega_s} I_2^2 \frac{R_2}{s} \tag{5.28}$$

Neglecting load and friction torques in the motor, we have

$$T = J\frac{d\omega}{dt} = -J\omega_s \frac{ds}{dt} \text{ since } \omega = \omega_s (1 - s) \tag{5.29}$$

The total energy loss in the rotor resistance of induction motor, when its slip changes from s_1 to s_2 is given as,

$$W = 3\int_{t_1}^{t_2} I_2^2\, R_2\, dt, \tag{5.30}$$

where t_1 and t_2 are the instants corresponding to s_1 and s_2 respectively.

Equating Eqns. (5.28) and (5.29), we have

$$3I_2^2 R_2 = -J\omega_s^2\, s\cdot\frac{ds}{dt}$$

Substituting this in Eqn. (5.30), we get

$$W = -J\,\omega_s^2 \int_{s_1}^{s_2} s\cdot ds$$

$$= \tfrac{1}{2} J\,\omega_s^2\,(s_1^2 - s_2^2) \text{ joules} \tag{5.31}$$

During starting, the slip of the induction motor changes from 1 to 0 and hence, the energy lost in the rotor circuit is given as

$$W_{st} = \frac{1}{2} J \omega_s^2 \text{ joules} \tag{5.32}$$

It may be noted that the energy lost in the rotor circuit does not depend on the rotor circuit resistance, but only on the moment of inertia of the rotating masses and the initial and final speeds. From Eqn. (5.32) it can be seen that the energy lost in the rotor circuit, during starting, is equal to the kinetic energy of the rotor at its final speed.

The above observation that the total energy loss in the rotor circuit during a change in speed is independent of the magnitude of the rotor resistance may be explained as follows. An increase in rotor resistance will reduce the magnitude of the current at a given speed and, hence, reduce the magnitude of the accelerating torque. This, in turn, will mean that the time taken to accelerate will be lengthened.

The fact that the total energy loss in the rotor circuit does not depend on the rotor circuit resistance doesn't imply that the total energy loss in the motor is independent of rotor circuit resistance. Neglecting the magnetizing current, both the stator current and rotor current referred to stator will be equal and, therefore,

$$\frac{\text{Stator copper losses}}{\text{Rotor copper losses}} = \frac{R_1}{R_2}.$$

Hence, although the energy lost in the rotor circuit is independent of R_2, the energy lost in the stator will be decreased by an increase in R_2, so that the total energy lost in the motor can be decreased by an increase in R_2.

An expression for the total energy lost in the motor during a change in speed can be easily obtained by modifying Eqn. (5.30) as

$$W_m = 3\int_{t_1}^{t_2} I_2^2 (R_1 + R_2)\, dt \tag{5.33}$$

Since $\qquad I_2 \simeq I_1$

Going through the same steps as from Eqn. (5.28) to Eqn. (5.31), we obtain

$$W_m = \frac{1}{2} J\omega_s^2 \left(1+\frac{R_1}{R_2}\right)(s_1^2 - s_2^2) \tag{5.34}$$

During starting, the energy lost in the motor will be,

$$(W_m)_{st} = \frac{1}{2} J\omega_s^2 \left(1+\frac{R_1}{R_2}\right) \tag{5.35}$$

It is obvious that the first term in the right hand side of the above equation represents the energy lost in the rotor circuit, which is equal to the kinetic energy of the rotor and the second term denotes the energy lost in the stator.

The energy relations derived, so far, were under the assumption that the load torque on the motor was absent. Let us, now, consider the presence of a load torque T_L. Under this condition

$$T_M - T_L = J\frac{d\omega}{dt}$$

i.e.,
$$dt = J.\frac{d\omega}{T_M - T_L} \tag{5.36}$$

Also,
$$d\omega = -\omega_s \cdot ds \tag{5.37}$$

Energy lost in the rotor circuit

$$W = 3\int_{t_1}^{t_2} I_2^2 R_2 \cdot dt$$

$$= \omega_s \int s\, T_M\, dt \tag{5.38}$$

Substituting Eqns. (5.36) and (5.37) in Eqn. (5.38), we get

$$W = -J\omega^2 \int_{s_1}^{s_2} \frac{T_M}{T_M - T_L} s \cdot ds$$

$$= -J\omega_s^2 \int_{s_1}^{s_2} \left[1 + \frac{T_L}{T_M - T_L}\right] s\, ds \tag{5.39}$$

Comparing Eqns. (5.32) and (5.39) it can be seen that the energy lost in rotor during starting of an induction motor on load will be greater than that on no load. In the above equation both T_M and T_L are functions of speed. If these functional relations are known, it is possible to integrate the expression in Eqn. (5.39) either analytically or graphically in order to determine the energy loss.

It may be recalled that Eqns. (5.32) and (5.39) were derived irrespective of the nature of the speed-torque characteristics of the motor and load as well as the method of starting the motor. While the energy loss on rotor during starting of the motor on no load is independent of the method by which it is started, the energy loss during starting on load will be minimum with direct-on-line start, since T_M the torque developed by the motor will be higher in magnitude at rated voltage. During reduced voltage starting ($V < V_{rated}$), the energy loss increases, in spite of the reduction in starting current. This can be clearly seen from Eqn. (5.39) if we substitute the new motor developed torque T'_M instead of T_M, bearing in mind that

$$T'_M = \left(\frac{V}{V_{rated}}\right)^2 \cdot T_M < T_M.$$

The entire energy loss during starting is dissipated as heat in the rotor and stator winding of squirrel cage motors and mostly in the external resistances introduced in the rotor circuit of slip-ring induction motors. Larger the magnitude of this external resistance, in relation to both the rotor and stator resistance, the cooler the motor will operate. But, as observed earlier, an increased resistance in the rotor circuit will increase the starting time.

Example 5-4. A 10 kW, 3-phase, 400 V, 20 A, 50 Hz, 960 rpm, 0.88 power factor squirrel cage induction motor drives a certain load. The total moment of inertia of the drive is 0.5 kg-m^2. (a) Determine the number of starts per minute that this drive can make under no load conditions without exceeding the total power dissipated in the motor under rated conditions. (b) How many starts per minute could be made if the rated speed were 1440 rpm, all other conditions remaining the same.

Assume a ratio of stator to rotor resistance of unity and neglect magnetizing current and rotational losses.

Solution: Input power $= \sqrt{3}\cdot 400\cdot 20\cdot 0.88$

$= 12193.6$ watts

Full load losses $= 12193.6 - 10000$

$= 2193.6$ watts

Energy lost per minute under full load conditions $= 2193.6\cdot 60$ joules

(a) Energy lost by the motor during starting under no load $= \frac{1}{2} J\omega_s^2 \left(1 + \frac{R_1}{R_2}\right)$

$$= \frac{1}{2}\cdot 0.5\cdot \left(\frac{2\pi\cdot 1000}{50}\right)^2\cdot 2 \text{ joules}$$

$= 5483$

Therefore, number of starts that can be made $= \frac{2193.6.60}{5483}$

$= 24.$

(b) Energy lost by the motor during starting on no load, if the rated speed is 1440 rpm

$$= \frac{1}{2}\cdot 0.5 \left(\frac{2\pi\cdot 1500}{60}\right)^2\cdot 2 \text{ joules}$$

$= 12337$ joules

So, the number of permissible starts $= \frac{2193.6\cdot 60}{12337}$

$= 10.67$, say 10.

5.5 Methods to Reduce the Energy Loss During Starting

The following methods are commonly used to reduce the loss in energy during starting of motors:

(i) Reducing the moment of inertia of the rotor;

(ii) Starting of dc shunt motors by smooth adjustment of applied voltage;

(iii) Starting of multispeed induction motors in discrete steps of speed; and

(iv) Starting of induction motors by smooth variation of supply frequency (V/f control).

5.5.1 Reducing the Moment of Inertia of Rotor

It is obvious from Eqns. (5.21), (5.25), (5.27), (5.32) and (5.39) that the total energy loss in motors during transient operation can be reduced by reducing the moment of inertia of the drive system. In order to achieve such a reduction, a single motor of certain power rating can be replaced by two motors of one-half of the rating. As a consequence of such a change, motors with a smaller diameter and hence reduced moment of inertia will be employed.

Yet another method is to use specially designed motors having large axial length. For the same power rating and speed of rotation, these motors will have smaller diameter than the general purpose motors.

5.5.2 Starting of dc Shunt Motors by Smooth Variation oj Applied Voltage

This method necessitates the presence of a variable direct voltage source. Smooth adjustment of applied voltage is equivalent to applying the voltage in a large number of small voltage steps. Let us assume, for easiness sake, that the motor is on no load and develops a constant torque so as to ensure uniform acceleration of the rotor. Starting in steps assumes the presence of a number of discrete steps in speed like ω_{01}, ω_{02}, ω_{03}, ... and the starting process to take place from 0 to ω_{01}, from ω_{01} to ω_{02}, from ω_{02} to ω_{03} and so on upto the no load speed of the motor ω_0. Since the speed of dc shunt motor is approximately directly proportional to applied voltage, the steps in speed will be of equal magnitude, if the voltage is varied in equal steps. Therefore, the loss in energy during starting with m equal steps of voltage can be expressed as,

$$W_{st} = m\left[\frac{1}{2}\,J\left(\frac{\omega_0}{m}\right)^2\right]$$

$$= \left[\frac{1}{2}\,J\,\omega_0^2\right]\frac{1}{m} \tag{5.40}$$

From Eqn. (5.40), it is obvious that larger the steps in voltage, less will be the energy loss during starting.

5.5.3 Starting of Multispeed Induction Motors in Discrete Steps

Pole-changing motors of two or more ratios of speeds can be started in discrete steps of speed. Since there is no change in the expression for energy loss during starting of an induction motor from that of a dc shunt motor, it can be easily argued that this method of starting of multispeed induction motor will also involve only less energy loss during starting.

5.5.4 Starting of Induction Motors by Smooth Variation of Supply Frequency

Just as in the case of dc shunt motors being started by smooth adjustment of applied voltage, this method will involve changes in speed in a very large number of steps. If the speed steps are equal in magnitude and a large number of steps in frequency are effected, the loss in energy during starting will be given by the expression

$$W_{st} = \frac{1}{m}\left[\frac{1}{2} J\omega_0^2\right]$$

where ω_0 is the final speed achieved by the motor.

PROBLEMS

1. A 220 V dc shunt motor having a sufficiently large armature resistance has the following speed-torque characteristics:

Speed (rpm)	1050	825	610	400	200	0
Torque (N-m)	0	2	4	6	8	10

If the motor drives a constant torque load of 5 N-m, determine

(i) the stable speed at which the motor operates,

(ii) the time taken to accelerate from rest to the operating speed, assuming the total moment of inertia of the motor and the load to be 0.5 kg-m^2, and

(iii) the energy lost during acceleration on load.

2. Two identical 600 V separately excited dc motors connected in parallel and supplied with rated voltage and rated excitation accelerate on no load from rest to 3000 rpm in 4 seconds. If both these motors on no load were connected in series and supplied with a total voltage of 600 V for a period of 2 seconds and then reconnected in parallel with a supply voltage of 600 V for another 2 seconds, the field excitations to the motors being kept at rated value under all conditions, determine

(i) the ratio of energy lost during acceleration from rest to full speed, under both conditions, and

(ii) the ratio of energy input to the armature, under both conditions.

3. Derive an expression for the time taken by a dc series motor driving a constant torque load to accelerate from a speed N_1 rpm to N_2 rpm. The motor developed torque may be assumed to be inversely proportional to the speed. Hence, determine the energy dissipated in the armature circuit during the acceleration of the motor from N_1 to N_2.

4. A 220 V dc series motor driving a constant load torque runs at 200 radians /sec and draws a current of 20 A from the supply. The total resistance of the armature and the field is 1 ohm. The moment of inertia of the motor together with the load is 5 kg-m^2. Calculate the total energy dissipated in the armature circuit, if the motor starts from rest and attains the steady state speed of 200 radians/sec within a time of 2.5 seconds.

5. A three-phase induction motor takes 1.5 times full load current when started by a star delta starter. Determine the autotransformer tapping for starting the motor so that the starting current should not exceed twice the full load current. Calculate also the starting torque in terms of full load torque corresponding to the above percentage of tapping of the autotransformer starter, given that the full load slip is 4 per cent.

6. A 15 kW, 400 V, 30 A, 950 rpm, 50 Hz, 3-phase, star connected, squirrel cage motor takes 6 times full load current and develops 1.8 times full load torque at standstill condition of the rotor with rated voltage being applied to the motor.

(i) What voltage must be applied to develop full load torque at starting ?

(ii) What current will be drawn by the motor at that voltage ?

(iii) If the voltage calculated in (i) is obtained by means of an autotransformer, what will be the line current ?

(iv) If the starting current in the line is limited to full load current by means of an autotransformer starter, what will be the starting torque as a percentage of full load torque? Neglect magnetizing current and stator leakage impedance drop.

7. If the torque developed by a motor could be expressed as $2500/n$ N-m where n is the revolutions per second (rps) and the load torque on the shaft as $n^2/30$ N-m, determine the time taken for the motor speed to reach the value of 25 rps from standstill. The moment of inertia of the motor and the load adds upto 5 kg-m^2.

8. A three-phase squirrel cage induction motor has a basic speed torque curve as given below :

p.u. speed	0	0.25	0.50	0.75	0.90	0.98	1
p.u. torque	2	1.50	1.70	2.25	1.75	1.00	0

The motor is driving a constant torque load of 1 p.u. The combined inertia of the motor and load is such that it requires 1.6 seconds to bring them to rated speed from rest, with a constant accelerating torque equal to rated torque. With the motor driving the load under normal steady state conditions, the voltage applied suddenly drops to 0.5 p.u., due to some trouble in the electrical supply system. It remains at this reduced voltage for 0.8 second and is then brought back to its rated value by clearing of the fault. Assume that the under voltage release on the motor does not function.

(a) Will the motor stop?

(b) If not, what will be the lowest speed attained by the motor ?

(c) Plot the motor speed-time curve from the beginning of the reduced voltage conditions until the steady state speed is almost reached for rated voltage.

9. A pole changing induction motor, having two speed in the ratio of 1:2 is to be started from rest on no load. Prove that, if the motor were accelerated to the maximum speed using the high speed winding the heating of the motor would be twice that which would be produced from acceleration from rest on the low speed winding and with further acceleration from one-half of the speed on the high speed winding.

10. A 5 kW, 3-phase 400 V, 50Hz, 960 rpm induction motor has a speed torque characteristic as given below:

Speed (rpm)	1000	960	900	750	500	250	0
Torque (N-m)	0	49.7	86.9	111.8	84.5	74.5	99.4

The motor is driving a constant torque load of 45 N-m.

(a) Calculate the acceleration time from rest to the operating speed, if the total inertia of the load and the motor is 2.2 kg-m^2.

(b) Determine the average power lost during acceleration.

Answers

1. (i) 500 rpm (ii) 20.5 secs (iii) 8744 J
2. (i) 0.5618 (ii) 0.7069
4. 21000 J
5. 0.667; 0.36
6. (i) 288.14 V (ii) 134.16 A (iii) 100 A (iv) 0.18
7. 4.85 secs
8. (a) No, (b) 0.566 p.u.
10. (a) 5.92 secs (b) 1876 W

Chapter 6

Electric Braking

Whenever an electric drive is disconnected from the supply, the speed of the driving motor gradually decreases and becomes zero. this natural process of braking, for reasons mentioned below, may not be, often, satisfactory. In many applications it may be necessary to provide a braking torque by artificial means—by mechanical brakes or electrodynamically.

The electrical method of providing a retarding torque has several advantages. For example, little maintenance is required, whereas mechanical brakes require adjustment and replacement of the brake linings; no dirt is produced, whereas the wear of mechanical brakes produces dust; the heat may be produced in a more convenient place, or, in some cases, a portion of the energy of the system may be returned back to the supply; electric braking is smooth without snatching. However, some forms of electric braking require epuipment of higher rating than required for motoring alone, while some other types require additional pieces of equipment. Hence, economic considerations have a considerable bearing on the use of electric braking. Also, electric braking, normally, cannot provide a holding torque.

Based on the purpose for which braking is employed, there are two forms of braking, viz., braking while bringing the drive to rest and braking while lowering loads. In the first type, the device used for braking absorbs the kinetic energy of the moving parts and in the second, it absorbs, in addition to the kinetic energy, potential energy, usually gravitational, which can drive the system at an excessively high speed.

Braking, while stopping, may be employed for any one of the following purposes:

(i) reducing the time taken to stop,

(ii) stopping exactly at specified points, for example, in lifts; sometimes such precise stops are necessary for reasons of safety,

(ii) feeding back, at least a portion of the power, to the supply network.

Braking, while lowering loads enables us to achieve any one of the following objectives:

(i) controlling the speed at which the load comes down and limiting it to a safe value,

(ii) feeding power back to the supply.

6.1 Types of Braking

There are three types of electric braking, all of which are applicable to the usual types of electric motor:

(1) Regenerative braking
(2) Rheostatic braking
(3) Plugging or reverse current braking.

Regenerative braking implies operating the motor as a generator, while it is still connected to the supply network. Mechanical energy is converted into electrical energy, part of which is returned to the supply. Rest of the energy is lost as heat in the windings and the bearings of the electrical machine. Regeneration does not, in most cases, involve any switching operation, unless it is required to change the speed at which it becomes effective. Most electrical machines pass smoothly from motoring to generating regime, when overdriven by the load.

Rheostatic braking implies operating the motor as a generator so that the mechanical energy is converted into electrical energy, which is dissipated as heat in the resistance of the machine winding or in resistors connected to them as an electrical load.

Plugging involves reconnecting the power supply to the motor so that it tends to drive in the opposite direction. It is obvious that, left to itself, the system will come to rest and then accelerate in the reverse direction. In case, it is required to bring the drive system to rest, it is necessary to include a relay to disconnect the supply exactly at the instant when the motor stops. This is the most inefficient technique of electric braking since, in addition to dissipating the electrical energy converted from the mechanical energy, in the resistances of the circuit, the electrical energy drawn from the supply is wasted.

6.2 Braking of dc Motors During Lowering of Loads

6.2.1 DC Shunt motor

When a dc shunt motor (or separately excited motor), used in a hoisting mechanism, is switched on, for lowering a load, the torque developed by the motor and that due to load torque act in unison to accelerate the motor. With increase in speed, the emf induced in the armature also increases and attains a value equal to the applied voltage, when the speed becomes the ideal no load speed. At this moment, the armature current and, hence, the electromagnetic torque becomes zero, so that the downward motion of the hoist is sustained only by the downward moving load. When the speed becomes greater than the ideal no load speed, $E > V$ and, therefore, the armature current becomes negative. The drive, then, acts as a generator and provides the braking torque. The drive attains a steady state speed when the braking torque developed by the motor is equal to the load torque. During this operation, the emf induced in the armature is directed against the applied voltage as it happens during motoring operation. The type of braking that occurs is regenerative braking, since power is fed back to the supply.

The speed-torque characteristics of the motor during motoring and braking operation are shown in Fig. 6.1. There are actually three characteristics—one being the inherent one, with no additional resistance in the armature circuit, and the other two being the modified ones, corresponding to increased values of resistance in the armature circuit. It is clear, from the figure that braking can occur only at speeds greater than the ideal no load speed of the motor.

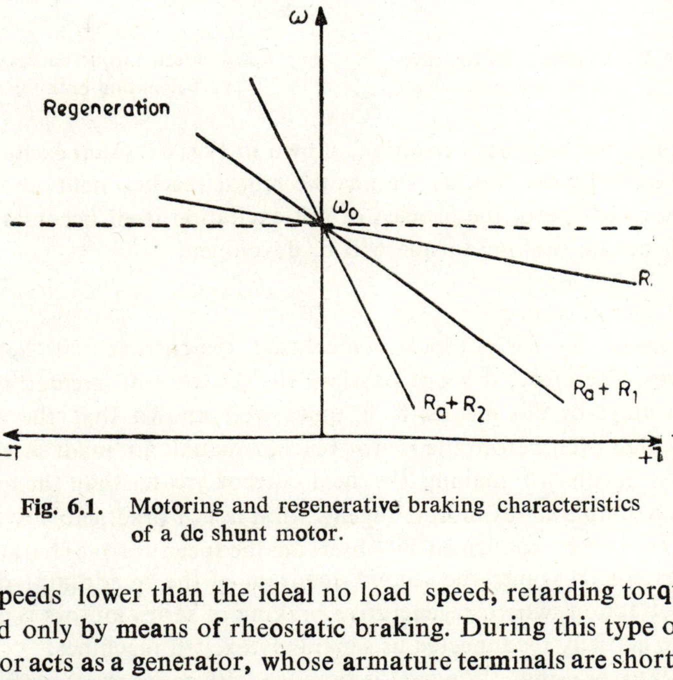

Fig. 6.1. Motoring and regenerative braking characteristics of a dc shunt motor.

At speeds lower than the ideal no load speed, retarding torque can be provided only by means of rheostatic braking. During this type of braking the motor acts as a generator, whose armature terminals are short circuited through an external resistance R_{ex} (Fig. 6.2). The voltage and speed equation corresponding to this operation are:

$$E = -I(R_a + R_{ex}), \tag{6.1}$$

and

$$\omega = -T \cdot \frac{(R_a + R_{ex})}{K_t K_e \phi^2}, \tag{6.2}$$

since $V = 0$

The speed-torque characteristics during rheostatic braking are as shown in Fig. 6.3; they approximate to straight lines passing through the origin. The slope of the characteristics depends on the total resistance in the armature circuit. Braking torque can be obtained at very low speeds, i.e., 0.1 to 0.07 of the rated speed. Obviously, complete stoppage of the downward moving load is not possible, because the current required for producing the braking torque is caused only by the movement of the load.

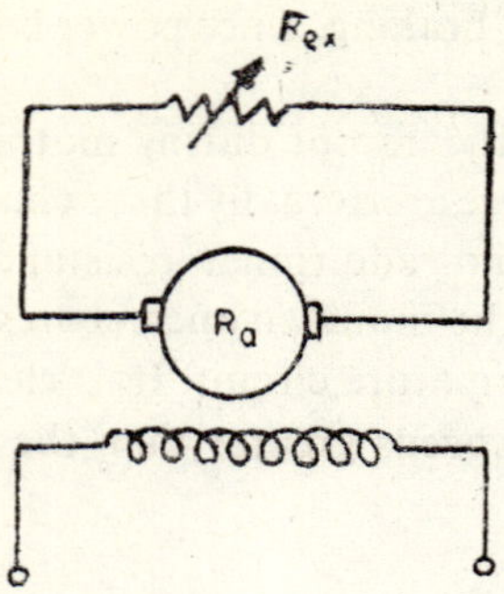

Fig. 6.2. Connections for rheostatic braking.

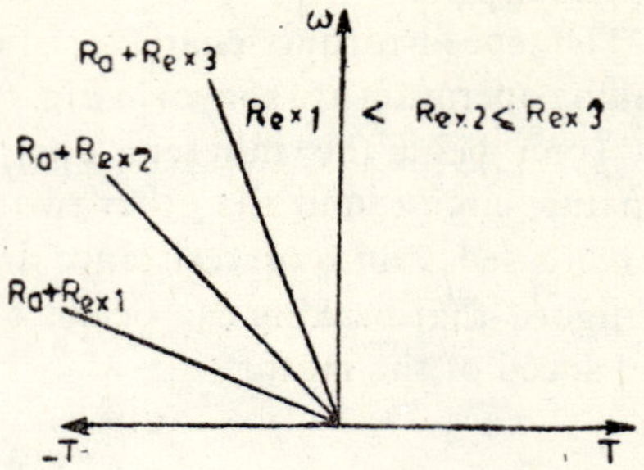

Fig. 6.3. Speed-torque curves during rheostating braking.

Instead of the separate excitation, shown in Fig. 6.2 shunt excitation can also be provided with. But, as soon as the motor reaches near about one-half of the rated speed, the process of self-excitation itself becomes ineffective and hence no braking torque will be developed.

6.2.2 DC Series Motor

If the speed of a series motor is increased, the current and, hence, the flux reduces. Therefore, it is not possible to get an emf greater than the terminal voltage by this means. It is quite well known that the speed is extremely high even before the motor reaches actual no load. Also, since there is no means of making the field current greater than the armature current in a simple series motor, regeneration is not practicable with series motors. This can be confirmed by observing the speed-torque characteristic, which does not cross into the second quadrant of the quadrantal diagram. In electric traction, where regenerative braking of series motors is used, the motors are actually reconnected as separately excited machines.

Rheostatic braking, of course, is possible with series motors (Fig. 6.4); but care should be taken to interconnect the armature winding and the field winding in such a way as to ensure that the direction of the current in the field remains the same, in spite of its change in the armature winding. Then only, the self-excitation of the series generator will take place. Also, the external resistance connected should not be of such a value to make the total resistance of the motor circuit greater than the critical value. Due to the presence of residual magnetism and the speed of rotation of the rotor, an emf is induced in the armature. Under its influence, current, flux and emf increases until they reach their steady state values, given by the equation :

$$E = I(R_a + R_f + R_{ex}), \tag{6.3}$$

where R_f is the resistance of the series field winding.

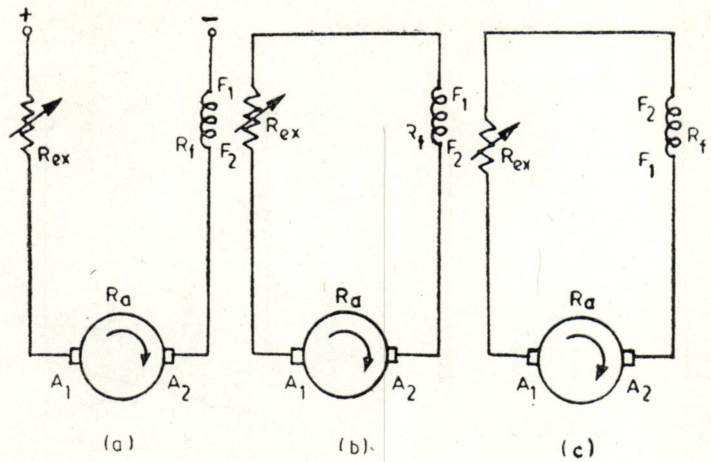

Fig. 6.4. Connections for rheostatic braking of dc series motor. (a) Motoring (b) Incorrect connections for braking (c) Correct connections for braking.

Rheostatic braking of series motors can also be effected by feeding their field winding from independent sources and connecting a resistance across the armature terminals, i.e., by reconnecting them as separately excited motors.

6.3 Braking while Stopping

6.3.1 DC Shunt Motor

Regenerative braking is possible, when an adjustable speed motor (a motor which operates at rated conditions with a weak field current), is used. Before bringing the motor to rest, the field excitation is increased to the permissible maximum value, as a result of, which, the speed of the motor falls to the minimum value and the kinetic energy released from the rotor is fed back to the supply. If the speed of the motor were to reduce in the ratio 4 : 1, the energy during braking would reduce in the ratio 16 : 1. Since the motor operates with a weak field at rated conditions, the armature has to be designed to carry a large current to produce the rated torque and, hence, the motor will be larger in size, poor in efficiency and costlier.

Plugging of dc motor involves reconnecting the motor to the line with reversed polarity; the motor now produces a torque in opposite direction to that of rotation. The rotor speed decreases until it becomes zero and then the rotor accelerates in the opposite direction. Therefore, plugging is used to get either a quick reversal or to get a rapid stop. For the latter, it is necessary to employ some means which will disconnect the motor at the time when it passes through zero speed.

The connections corresponding to the normal working and plugging of dc shunt motor are as shown in Fig. 6.5.

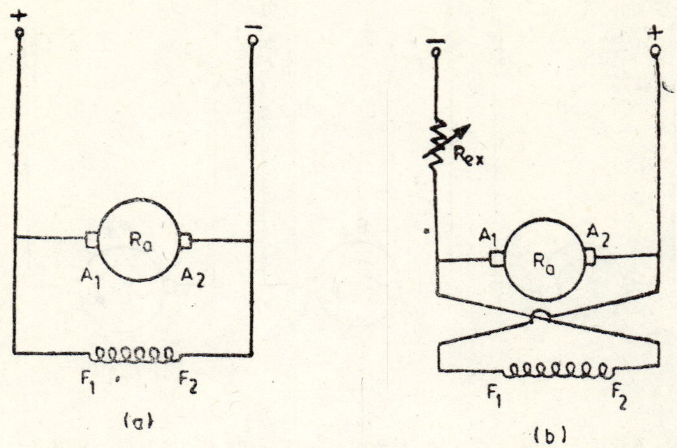

Fig. 6.5. Connections during plugging of dc shunt motor. (a) Motoring (b) Braking

In order to reverse the torque developed, it is necessary to reverse the magnetic field or the armature current. Since the field winding, usually, has a large time constant due to large inductance, armature current is reversed. When the armature current is reversed, the back emf of the armature will no longer oppose the applied voltage, but will aid it. This condition is actually worse than starting motor direct on line, since nearly double the supply voltage is applied across the armature having a low resistance. Therefore, additional resistance R_{ex} is inserted in the armature circuit, simultaneously with the reversing of connections of the armature.

The plugging operation of a dc shunt motor can be explained with the help of the quadrantal diagram shown in Fig. 6.6. The characteristic A is exactly of the same shape as the one shown in Fig. 3.9 for a dc shunt motor, with a sufficiently high series resistance in the armature, running in the reverse direction. The curve B corresponds to the operation of the motor in the forward direction, with no series resistance in the armature. For different values of constant load torques acting on the motor, different plugging torques are obtained. It can be observed from the quadrantal diagram that (a) the value of the plugging torque developed will be corresponding to the speed of operation of the motor before plugging, (b) the plugging torques vary only slightly with the motor load, since the speed of the shunt motor cannot change much with different loads, and (c) the plugging torque decreases to about one-half of its value corresponding to rated speed of the motor, at standstill conditions. It should be obvious, from Fig. 3.9, that the plugging torque can be varied by changing the magnitude of the resistance inserted in the armature circuit.

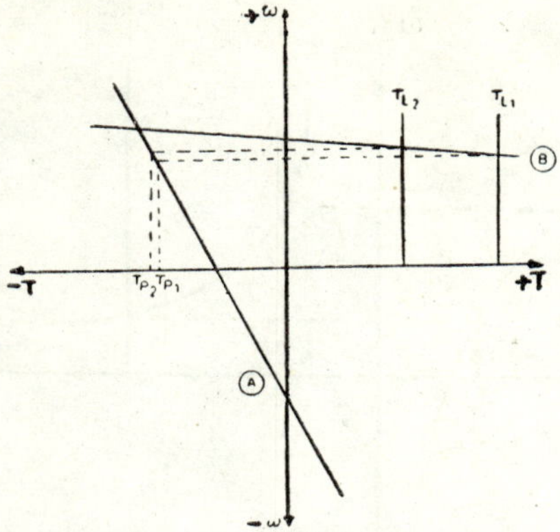

Fig. 6.6. Quadrantal diagram during plugging of dc shunt motor.

6.3.2 DC Series Motor

Rheostatic braking of dc series motors can be employed to bring the motor to rest, if the conditions necessary to cause self-excitation (as explained in section 6.2.2) are ensured.

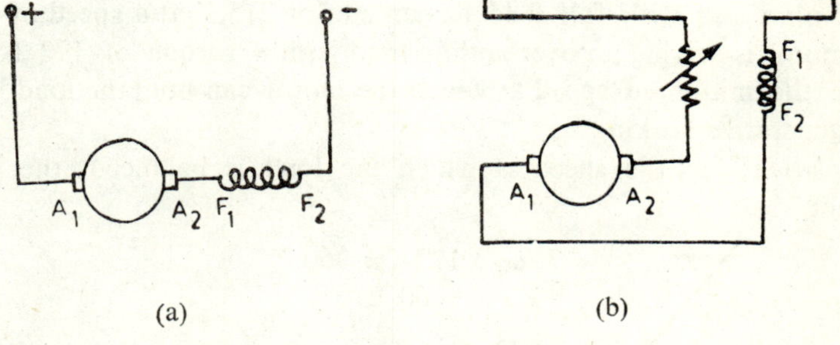

Fig. 6.7. Connections during plugging of dc series motor.
(a) Motoring (b) Braking

Figure 6.7 shows the circuit diagram during plugging of dc series motor. The quadrantal diagram during plugging is shown in Fig. 6.8. Curve A in this figure is of the same shape as the one shown in Fig. 3.13 for a dc series motor, with additional high resistance in the armature circuit, while operating in the reverse direction. It can be seen that the plugging torque reduces somewhat more rapidly with speed than in the dc shunt motor. The speed torque characteristic of the same motor, without having any extra resistance in the motor circuit, running in the forward direction, is depicted by curve B. For various constant torque loads acting on the motor, various magnitudes of plugging torque will be developed. In contrast to that of dc shunt motor, it can be seen that plugging torques vary significantly with

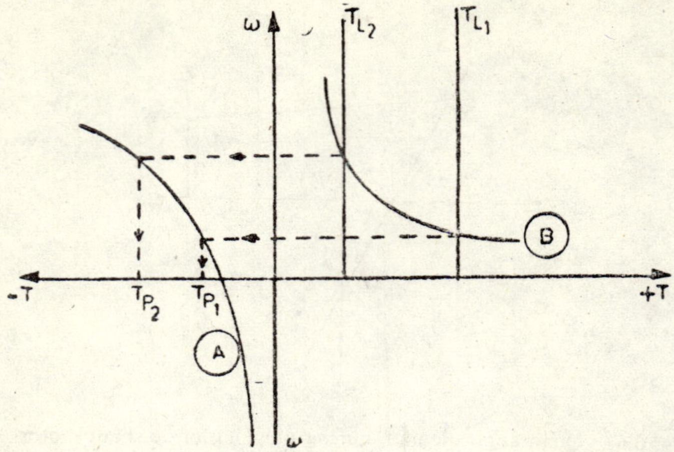

Fig. 6.8. Quadrantal diagram during plugging of dc series motor.

changes in load torque, that is, with changes in motor speed before plugging.

Example 6-1. A 220 V dc shunt motor has an armature resistance of 0.062 ohm and with full field has an emf of 215 V at a speed of 960 rpm. The motor is driving an overhauling load with a torque of 172 N-m. Calculate the minimum speed at which the motor can hold the load by means of regenerative braking.

Solution: At the speed at which the load is balanced, the armature current

$$I_a = \frac{T \cdot \omega}{E} = \frac{172 \cdot 2\pi \cdot 960}{60 \cdot 215}$$

$$= 80.42 \text{ A}$$

EMF induced during regenerative braking

$$E_g = V + I_a R_a$$

$$= 220 + 80.42 \cdot 0.062$$

$$= 225 \text{ V}$$

E is directly proportional to speed, being a shunt machine.

$$\text{Hence, speed} = \frac{960}{215} \cdot 225$$

$$= 1004.65 \text{ rpm.}$$

Example 6-2. A dc series motor is subjected to rheostatic braking against a load torque of 318.3 N-m. Determine the value of the resistance to be connected in the motor circuit to limit the speed to 480 rpm. The total resistance of the armature and the field is 0.24 ohm and the magnetization curve corresponding to 900 rpm is as given below:

Field current (A)	20	40	60	80	100
EMF (V)	261	540	738	882	945

Neglect rotational losses of the motor.

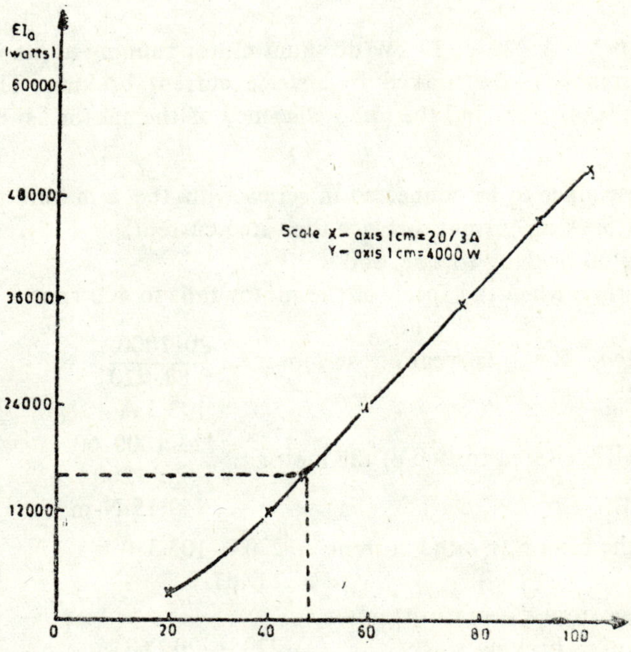

Fig. 6.9. Plot of EI_a vs I_a for Example 6-2.

Solution: Input power to machine during rheostatic braking

$$= \frac{2\pi N}{60} \cdot T$$

$$= 2\pi \cdot \frac{480}{60} \cdot 318.3$$

$$= 16000 \text{ W}$$

Since rotational losses are negligible, this must correspond to the electromagnetic power developed by the generator, i.e., the product $E.I_a$. In order to determine E and I_a separately, let us plot a graph of EI_a against I_a of the machine at the operating speed of 480 rpm, using the table below:

I_a	20	40	60	80	100
E	139.2	288	393.6	470.4	504
EI_a	2784	11520	23616	37632	50400

I_a corresponding to $EI_a = 16000$ W is found to be $= 48$ A
Corresponding $E = 333.3$ V

$$\text{Hence, total resistance of circuit} = \frac{E}{I_a} = 6.94 \text{ ohm.}$$

External resistance added in the motor circuit $= 6.70$ ohm.

Example 6-3. A 220 V, 20 kW dc shunt motor running at its rated speed of 1200 rpm is to be braked by reverse current braking. The armature resistance is 0.1 ohm and the rated efficiency of the motor is 88 per cent. Calculate

(a) the resistance to be connected in series with the armature to limit the initial braking current to twice the rated current,
(b) the initial braking torque, and
(c) the torque when the speed of the motor falls to 400 rpm.

Solution: Rated current of the motor $= \dfrac{20 \cdot 1000}{0.88 \cdot 220} = 103.3$ A

Rated torque of the motor $= \dfrac{20 \cdot 1000 \cdot 60}{2\pi \cdot 1200} = 159.15$ N-m

(a) E of the motor at rated current $= 220 - 103.3 \cdot 0.1 = 209.67$ V

Voltage across armature circuit as soon as braking starts $= 220 + 209.67 = 429.67$ V

Braking current at the beginning of braking $= 2 \cdot 103.3$ A

Hence, resistance in the armature circuit $= \dfrac{429.67}{206.6} = 2.08$ ohm

Therefore, extra resistance to be added to the motor armature $= 2.08 - 0.1 = 1.98$ ohm.

(b) Being a shunt motor, the flux per pole will be constant. Hence, torque $\propto$ current. Therefore, initial braking torque will be twice the full load torque,
i.e., $= 2 \cdot 159.15 = 318.30$ N-m.

(c) E at 400 rpm $= \dfrac{209.67}{1200} \cdot 400 = 69.89$ V

Hence, braking current at 400 rpm $= \dfrac{220 + 69.89}{2.08}$

$= 139.37$ A

Braking torque $= \dfrac{159.15}{103.3} \cdot 139.37$

$= 214.72$ N-m.

6.4 Electric Braking of Induction Motors

6.4.1 Regenerative Braking

Regenerative braking can take place only when the rotor rotates in the same direction as that of the stator magnetic field, but with a speed greater than the synchronous speed. Such a state occurs during any one of the following processes:

(i) Switching over to a low frequency supply in frequency controlled induction motors in order to reduce the speed of operation of the drive.
(ii) Downward motion of a loaded hoisting mechanism.
(iii) Switching over to a larger pole number operation from a smaller one in multispeed squirrel cage motors.

In all the above cases, the slip and the torque developed become negative and, thus, the machine acts as a generator, receiving mechanical energy and giving it back to the supply as electrical energy.

During switching over from smaller pole number to larger or from high frequency to low frequency of supply, the change in speed from a higher

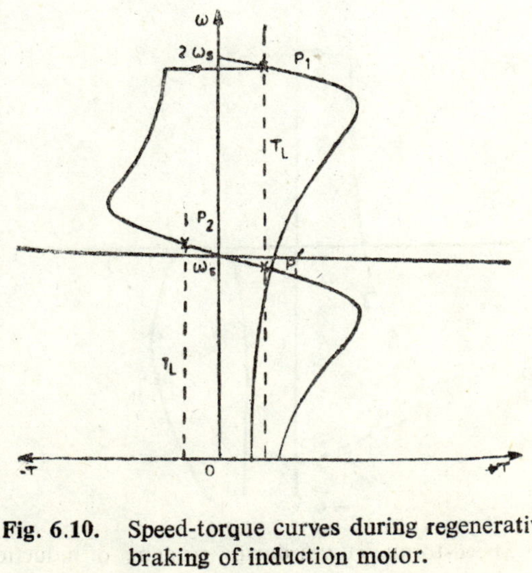

Fig. 6.10. Speed-torque curves during regenerative braking of induction motor.

value to a lower one takes place as per the speed-torque curve shown in Fig. 6.10 (refer to points P_1 and P_1').

If, however, the load drives the motor above synchronous speed, no switching operation is required. Once the machine is driven above synchronous speed, the braking operation automatically starts. The operating point will be decided by the magnitude of the load torque and the nature of the speed-torque of the machine during generator operation (see point P_2 in Fig. 6.10). By varying the resistance in the rotor circuit, it is possible to operate at any speed higher than the synchronous speed, during braking.

If the driving torque of the load exceeds the maximum braking torque of which the motor is capable, the system will be unstable and the speed will rise still further, probably to a disastrous value, since, the faster the motor runs, the less will be the braking torque developed.

6.4.2 Plugging or Reverse Current Braking

Plugging occurs when the rotor and the stator magnetic field move in opposite directions. Such an operation takes place either during reversal of the direction of rotation of the rotor, while running or when a negative torque is applied to the shaft of the motor. In both cases, the slip becomes greater than unity, since

$$s = \frac{\omega_s - (-\omega_r)}{\omega_s} = \frac{\omega_s + \omega_r}{\omega_s} = 1 + \frac{\omega_r}{\omega_s} \tag{6.4}$$

The speed-torque characteristics during plugging are shown in Fig. 6.11. They are actually extensions of the motor characteristics in the second and fourth quadrant. This method of braking is employed in hoisting mechanisms for reducing the speed of a downward moving load. A stable

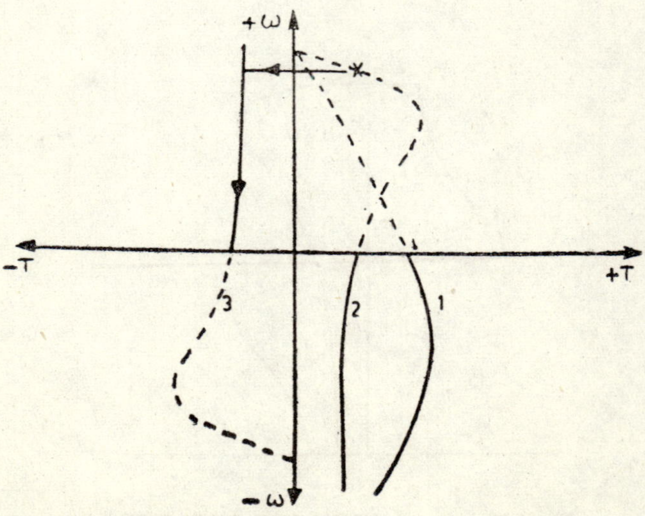

Fig. 6.11. Speed-torque curves during plugging of induction motor.

operation will be obtained only when the condition for steady state stability $\left(\frac{dT_L}{d\omega} > \frac{dT_M}{d\omega}\right)$ is satisfied. From curves 1 and 2 of Fig. 6.11 it is clear that this will happen only when the rotor resistance is sufficiently large. By varying this resistance, it is possible to change the speed during braking. Addition of rotor resistance has another advantage of reducing stator current.

Reverse current braking is obtained by simply reversing the phase sequence of the supply to the motor stator, i.e., just by interchanging any two supply lines. For bringing the motor to rest, it is necessary to disconnect the supply at zero speed; otherwise, reversal of direction of rotation will take place (refer to curve 3 of Fig. 6.11).

During plugging, both the electrical energy consumed from the supply and the mechanical energy applied at the shaft, are spent as heat developed in the windings of the motor together with any external resistances connected in series with the rotor.

6.4.3 DC Rheostatic Braking

While motoring the stator magnetic field of the induction machine rotates at synchronous speed in the same direction as that of the rotor, but slightly faster than the rotor conductors. If the stator windings were disconnected from the ac supply and fed with dc, the magnetic field produced will be stationary in space, thus, making the rotor conductors move past the field with a speed $(1-s)\omega_s = S\omega_s$. The currents induced in the rotor conductors will be opposite in direction to that corresponding to motoring operation, producing a braking torque.

Although the airgap flux is stationary, there will be the same number of poles as when the machine is excited with ac and the rotor currents will, therefore, have a frequency which is near to the frequency corresponding to synchronous speed $\left(=\frac{\omega_s}{2\pi}\right)$ initially, but which decreases with rotor speed and becomes zero at standstill, i.e., the frequency of the rotor current can be expressed as Sf, where $f = \omega_s/2\pi$. Likewise, the induced rotor emf decreases from a maximum when the motor is running, to zero at standstill. Or, in other words, the emf induced in the rotor will be given by SE_1, when E_1 is the magnitude of the emf induced in the rotor when it rotates past the field at a speed ω_s. It is, thus seen that the conditions in the rotor during dc rheostatic braking with the speed falling from synchronous to standstill are very much the same as when the motor accelerates in the normal manner. Hence, the equivalent circuit of the rotor can be represented as shown in Fig. 6.12(a). Dividing the voltage and the impedance by S, we get the circuit shown in Fig. 6.12(b) carrying the same current I_2.

Since the stator winding carries only dc current, the inductance of the stator has no effect under steady state operation. The dc voltage applied

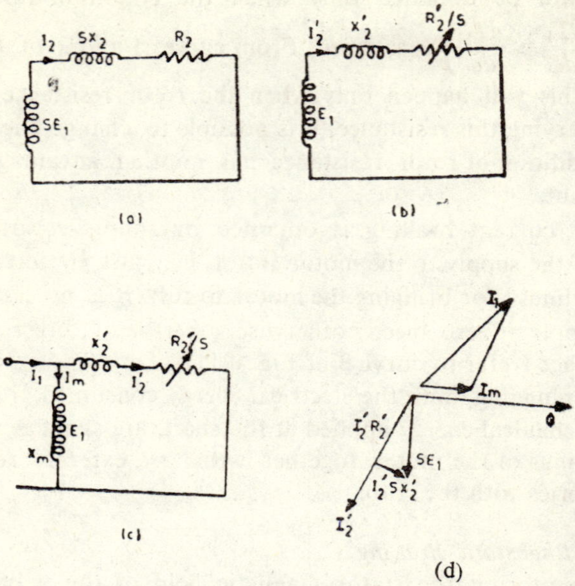

Fig. 6.12. DC rheostatic braking of induction motor.
(a) Equivalent circuit of rotor
(b) Referred equivalent circuit
(c) Referred equivalent circuit for braking operation
(d) Corresponding phasor diagram.

across the stator winding is fixed only by the value of the stator winding resistance. Note, also, that there would not be any stator core losses. But, the rotor core losses will be of considerable magnitude and hence the rotor resistance value referred to stator, which determines the torque developed during braking, must be corrected to take this into account. The emf induced in the rotor, at synchronous speed, E_1 will be given by $I_m X_m$, where I_m represents the magnetizing current passing through the magnetizing reactance X_m. The circuit in Fig. 6.12(b) can be transformed into as shown in Fig. 6.12(c). The phasor diagram corresponding to Fig. 6.12(c) is depicted in Fig. 6.12(d).

It may be noted that the rotor current is alternating in nature, although it is created by a constant flux, caused by a magnetizing mmf $I_m N_1$, which remains stationary in space. However, as viewed from the rotor, both the magnetizing mmf and the resulting flux in the stator will appear as alternating. Also, since the rotor ampere turns $I_2 N_2$ $(= I_2' N_1)$ must be balanced by the stator ampere turns so as to maintain the magnetizing ampere turns $I_m N_1$, the stator ampere turns $I_1 N_1$, as seen from the rotor, represents the vector sum of $I_m N_1$ and $I_2' N_1$. The effective ac I_1 depends on the magnitude of dc carried by the stator and the nature of stator winding connections, as explained in the next section.

From what has been said above, it is clear that the operation of the induction motor during dc rheostatic braking can be studied by using the equivalent circuit (the type of which we are already familiar) shown in Fig. 6.12(c).

The braking torque can be found from the expression

$$T_b = \frac{3}{\omega_s} \cdot I_2'^2 \frac{R_2'}{S} \text{ N-m}$$

$$= \frac{3}{\omega_s} I_1^2 \cdot \frac{x_m^2 \left(\frac{R_2'}{S}\right)}{\left(\frac{R_2'}{S}\right)^2 + (x_m + X_2')^2} \text{ N-m} \qquad (6.5)$$

The above equation enables us to determine the braking torque at any speed of a given induction motor, when excited by direct current. I_1 corresponds to an equivalent ac, which would have produced the same mmf as that by the actual dc carried by the stator windings.

Differentiating Eqn. (6.5) with respect to S and equating it to zero for a maximum, we get

$$T_{b(max)} = \frac{3}{\omega_s} \cdot I_1^2 \frac{x_m^2}{2(x_m + X_2')} \qquad (6.6)$$

The maximum braking torque occurs at a speed,

$$S = \frac{R_2'}{x_m + X_2'} \qquad (6.7)$$

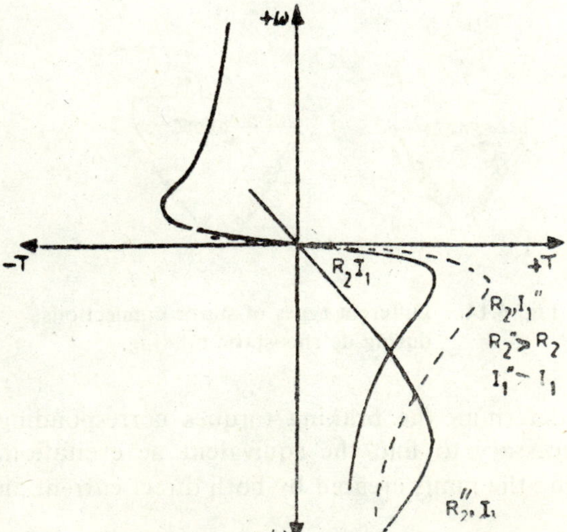

Fig. 6.13. Speed-torque curves during dc rheostatic braking.

The speed-torque characteristics during dc rheostatic braking are as shown in Fig. 6.13. Effect of variation of I_1 and R_2 are also indicated in the same figure. With increasing R_2, the speed at which maximum braking torque occurs also increases. But, the maximum braking torque would not increase in proportion to the square of the current, as pointed out by Eqn. (6.6), since the value of x_m decreases due to saturation caused by increase in I_1.

This method (dc rheostatic braking of induction motors) is quite convenient in practice and commonly used with active loads. It ensures complete and effective control, if corresponding to the speed of operation, I_1 or R_2 or both are adjusted to maintain the necessary braking torque. The advent of automatic control of rheostatic braking of induction motors using closed loop systems, has made the use of induction motors more popular than dc motors, especially for drives employed in mine hoists.

(i) *Different methods of feeding dc to stator winding*: There are a number of ways in which the stator winding can be connected as dc field winding, but four widely used connections are as shown in Fig. 6.14.

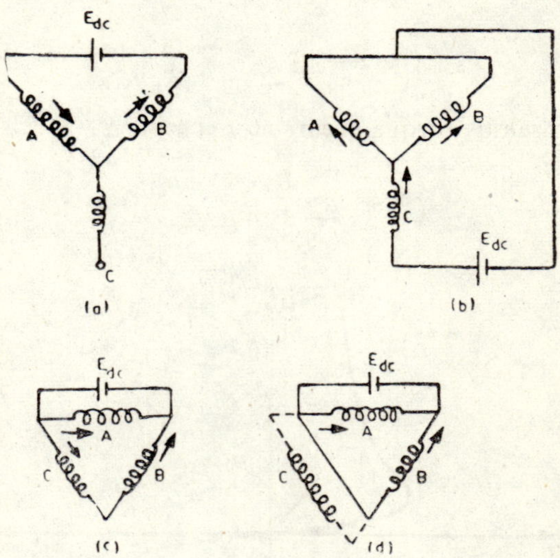

Fig. 6.14. Different types of stator connections during dc rheostatic braking.

In order to determine the braking torques corresponding to different speeds, it is necessary to find the equivalent ac excitation. This can be found by equating the mmfs created by both direct current and alternating current.

Let us first, consider the connection shown in Fig. 6.14(a). Phases A and B carry equal direct currents, but in opposite direction. Phase C doesn't carry any current at all. Assuming that the space mmfs produced by the dc flowing in phases A and B are sinusoidal in nature (or neglecting space harmonic mmfs), the resultant mmf can be determined as shown in Fig. 6.15.

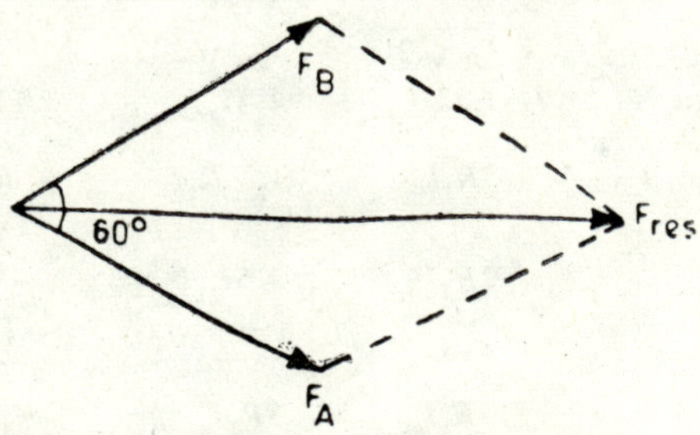

Fig. 6.15. Space phasor diagrams of mmfs produced with connection of Fig. 6.14 (a).

$$\text{Resultant mmf } F_{res} = 2I_{dc}N_{eff}\cos 30°$$
$$= \sqrt{3}\, I_{dc}N_{eff}, \tag{6.8}$$

where N_{eff} is the effective turns per phase.

$$\text{Mmf due to } I_{ac(rms)} \text{ flowing in all the three phases} = \frac{3}{2}(\sqrt{2}\, I_{ae})\, N_{eff} \tag{6.9}$$

Equating both mmf's, we have

$$I_{ac} = \sqrt{2/3}\, I_{dc} \tag{6.10}$$

The required magnitude of direct voltage V_{dc} to feed a current I_{dc} is given by

$$V_{dc} = I_{dc} \times 2R_1,$$

where R_1 denotes the stator resistance per phase.

Power spent on excitation

$$P_{exc.} = V_{dc}I_{dc} = 2R_1I_{dc}^2$$
$$= 3R_1I_{ac}^2 \tag{6.11}$$

The corresponding values for different methods of connection are given in Table 6.1. In all the above schemes, for the same values of the equivalent ac, same magnetic fields and hence same magnitudes of braking torques are obtained and the excitation power for all the connections also remain the same.

From Table 6.1, it is clear that the connection shown in Fig. 6.14(a) is the most preferred one, since that demands the least current at maximum voltage.

Table 6.1. Currents, voltage and excitation power corresponding to different connections of stator winding during dc rheostatic braking

	Schemes shown in Fig. 6.14			
	a	b	c	d
Equivalent rms ac, I_{ac}	$I_{dc}\sqrt{2/3}$	$I_{dc}/\sqrt{2}$	$I_{dc}(\sqrt{2/3})$	$I_{dc}/\sqrt{6}$
DC excitation current, I_{dc}	1.22 I_{ac}	1.41 I_{ac}	2.12 I_{ac}	2·45 I_{ac}
DC excitation voltage, V_{dc}	$2R_1I_{dc}$	$1.5\,R_1I_{dc}$	$\frac{2}{3}\,R_1I_{dc}$	$\frac{1}{2}\,R_1I_{dc}$
Excitation Power, P_{exc}	$2R_1I_{dc}^2 = 3R_1I_{ac}^2$	$1.5R_1I_{dc}^2 = 3R_1I_{ac}^2$	$\frac{2}{3}\,R_1I_{dc}^2 = 3R_1I_{ac}^2$	$\frac{1}{2}\,R_1I_{dc}^2 = 3R_1I_{ac}^2$

Example 6-4. A 3-phase, 440 V, 50 Hz, 10 pole star connected induction motor has the following parameters : $R_1 = 0.15$ ohm, $R_2 = 0.45$ ohm, $X_1 = 0.6$ ohm, $X_2 = 1.8$ ohms, $s_f = 0.05$ and the ratio of effective stator to rotor turns $1/\sqrt{3}$. The motor is to be braked at rated speed and an external resistance of 1.75 ohms per phase (referred to stator) has been inserted into the rotor circuit. Determine the initial braking torque for the following two cases of braking:

(i) dc rheostatic braking,

(ii) reverse current braking.

Solution: Rotor resistance referred to stator $= 0.45\left(\frac{1}{\sqrt{3}}\right)$

$= 0.15$ ohm

Rotor leakage reactance of motor referred to stator

$$= 1.8\left(\frac{1}{\sqrt{3}}\right)^2$$

$= 0.6$ ohm.

(i) The equivalent circuit of the motor corresponding to a slip of 0.05 and the equivalent circuit of the machine during rheostatic braking at rated speed are shown in Fig. 6.16(a) and (b). The magnitude of emf E_1 will be given by

$$E_1 = \sqrt{\frac{3^2 + (0.6)^2}{(3.15)^2 + (1.2)^2}}\cdot\frac{440}{\sqrt{3}}\text{ V} = 230.56\text{ V}$$

During rheostatic braking $S = 1 - 0.05 = 0.95$

$$I_2 = \frac{230.56}{\sqrt{(0.6)^2 + 2^2}} = 110.42\text{ A}$$

Initial braking torque is given by

$$\frac{3I_2^2R_2}{S\omega_s} = \frac{3\,(110.42)^2\cdot 2.60}{0.95\cdot 2\pi\cdot 600}$$

$$= 1225.58\text{ N-m.}$$

(ii) During reverse current braking initial slip will be

$$= 2 - 0.05 = 1.95$$

Therefore, initial current

$$I_2 = \frac{440}{\sqrt{3}\sqrt{\left[0.15 + \left(\frac{1.9}{1.95}\right)\right]^2 + (1.2)^2}}$$

$$= 159.42 \text{ A}$$

$$\text{Initial braking torque} = \frac{3\,(159.42)^2 \cdot 1.9/1.95}{2\pi \cdot 600/60}$$

$$= 1182.35 \text{ N-m.}$$

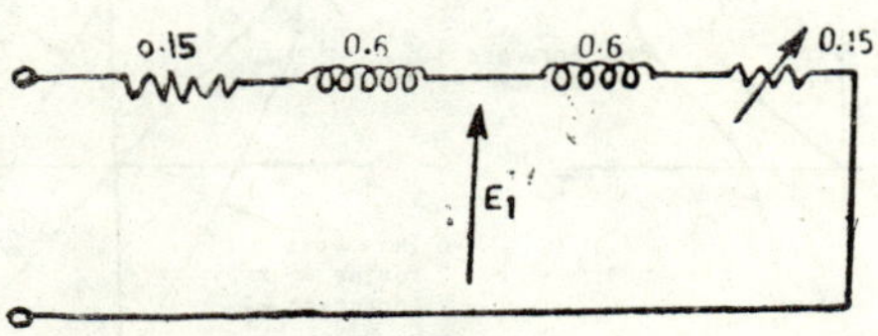

(a)

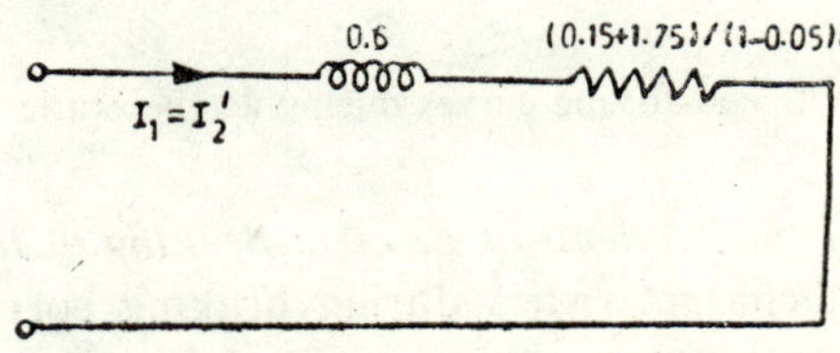

(b)

Fig. 6.16. Equivalent circuit:
(a) during motoring;
(b) during dc rheostatic braking.

6.4.4 *AC Rheostatic Braking*

AC rheostatic braking is effected by feeding single-phase supply to properly connected stator windings (as in the case of dc rheostatic braking). A little thinking shows that this condition of operation is in fact identical with the operation of a three-phase induction motor on a single phase supply. A doubt arises immediately as to how the same condition results both in motoring and braking torque. The answer lies in clearly understanding how and under what condition motoring torque is developed under single phase operation of three phase induction motor.

Using revolving field theory, the torques produced by the forward field, backward field and the net torque developed in a single-phase induction motor can be determined as shown in Fig. 6.17. It can be seen from the figure that the net torque of single-phase motor becomes zero at little less than synchronous speed. In between this speed and the synchronous speed the net torque developed is negative. This zone of negative torque can be widened by increasing the rotor circuit resistance to such a value as to get negative torque developed for all speeds from synchronous to zero (refer to Fig. 6.17). Since this resistance is much more than the rotor resistance of a well designed squirrel cage induction motor, ac rheostatic braking can be used, normally with slipring induction motors, in which external resistances can be inserted in the rotor circuit.

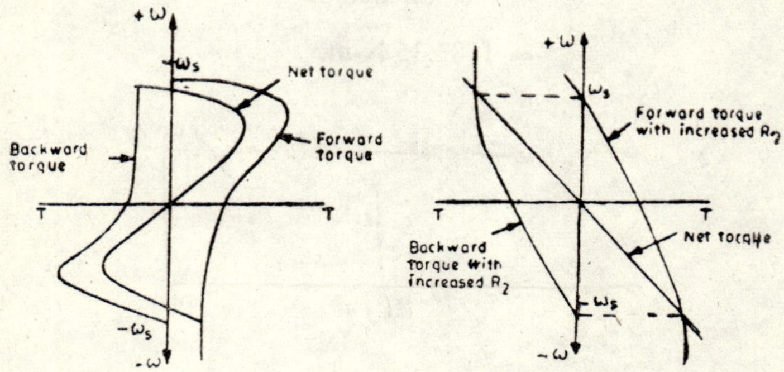

Fig. 6.17. Speed-torque curves during ac rheostatic braking.

6.4.5 *Comparison of the Methods of Electric Braking of Induction Motors*

The speed-torque characteristics during braking corresponding to the various methods of braking are shown in Fig. 6.18.

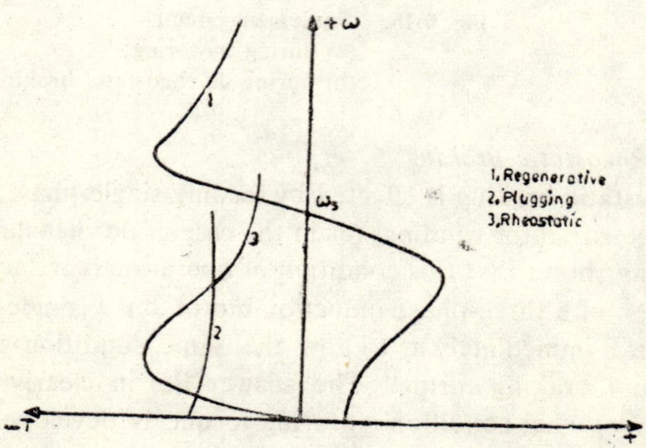

Fig. 6.18. Speed-torque curves of induction motor during different methods of braking.

Plugging or reverse current braking has the following advantages:

(i) Simple control scheme, except for the requirement of a relay to stop the motor, and
(ii) Uniform current loading of all the three phases during the process of braking.

Considerable loss of energy as well as increased heating of the machine, possibility of the motor running in the reverse direction in case of malfunctioning of the relay used to stop the motor and appearance of quite high voltages at the sliprings are the disadvantages of plugging.

DC rheostatic braking is the most preferred method as regards losses and heating of the machine. Another merit of this method is that braking torque can be developed even at low values of speed and that there will be no risk of reversal of the direction of rotation of the motor. It is possible, by passing very high exciting currents for very short time, to bring the machine to rest very quickly in an emergency. Yet another major advantage is the possibility of having automatic control of the braking process by means of closed loop.

Regenerative braking has the disadvantage of the possibility of braking only at high speeds (super synchronous speeds). This method can be used only in hoisting type of mechanisms or with a multispeed squirrel cage motor.

6.5 Braking of Synchronous Motors

In principle, a synchronous motor may be stopped by either plugging or rheostatic braking. But, the former method has certain disadvantages. Synchronous motors usually, are of large power rating and the line disturbances that occur during plugging are beyond the permissible limits. Also, since the plugging torques arise from the damper winding and unless it is specially designed for the purpose, the torques developed during plugging are smaller in magnitude than those obtained by rheostatic braking. Hence, rheostatic braking is almost exclusively used for braking synchronous motors.

6.5.1 Rheostatic Braking of Synchronous Motors

A synchronous motor is stopped by rheostatic braking by simply disconnecting it from the supply lines and reconnecting the machine to a balanced three-phase resistor, keeping the field excitation constant. In case greater braking torques are required, the field excitation can be increased. During braking the machine acts almost as a short circuited generator, except that the terminal voltage and frequency decrease with speed and both become zero at standstill.

The braking current at any speed ω of the rotor is given by

$$I_{br} = \frac{\omega L_{af} I_f}{\sqrt{2}\sqrt{r^2 + (\omega L_s)^2}}, \tag{6.12}$$

where L_{af}, I_f, r and L_s denote the mutual inductance/phase between field and stator winding, field current, magnitude of braking resistance/phase including stator resistance and synchronous inductance/phase respectively. $\frac{\omega L_{af} I_f}{\sqrt{2}}$ actually represents E_{rms}, the induced emf per phase in the stator due to the field current I_f and rotational speed ω. Since the resistance r will be small compared with X_s, Eqn. (6.12) can be simplified as,

$$I_{br} = \frac{L_{af}}{\sqrt{2}\,L_s} \cdot I_f = K_1 I_f, \tag{6.13}$$

where K_1 is a constant for the machine.

Therefore, if excitation current is kept constant, the braking current in the stator winding remains constant at down to very low speeds. I^2R losses generated or power made available for braking, hence, remains more or less constant throughout the braking process, i.e., the braking torque T_{br} produced by these losses in the braking resistances, increases as the speed decreases. The speed at which maximum braking torque is developed can be determined by differentiating the torque expression with respect to ω and equating it to zero, as follows:

$$T_{br} = \frac{P_{br}}{\omega} = \frac{3 I_{br}^2 r}{\omega}$$

$$= \frac{3 \cdot L_{af}^2 I_f^2 r}{2} \cdot \frac{\omega}{r^2 + \omega^2 L_s^2} \tag{6.14}$$

$$= K_2 \cdot \frac{\omega}{r^2 + \omega^2 L_s^2}, \tag{6.15}$$

where K_2 is a constant for the machine.

Differentiating Eqn. (6.15) for a maximum, we get

$$\omega_{\max T\,br} = \frac{r}{L_s} \tag{6.16}$$

i.e., the maximum braking torque occurs at approximately that speed for which the machine reactance equals the value of the braking resistance. The appearance of high torque immediately prior to stopping has a very favourable effect on the braking process.

Example 6·5. A 3000 kW, unity power factor 3-phase star connected, 2300 V, 30 pole, 50 Hz synchronous motor has a stator resistance and synchronous reactance of 0.2 and 2 ohms per phase respectively. Compute the magnitude of the stator resistance per phase to be added to brake the motor so that the initial braking current does not exceed the rated current of the motor, if the field excitation is kept constant at the value which would result in unity power factor at rated load.

Also, determine the initial braking torque.

Solution: Rated KVA/Phase $= \dfrac{3000}{3} = 1000$

Rated voltage/phase $= \dfrac{2300}{\sqrt{3}} = 1330$

Rated current/phase $= \dfrac{1000 \cdot 1000}{1330}$

$= 751.88$ A.

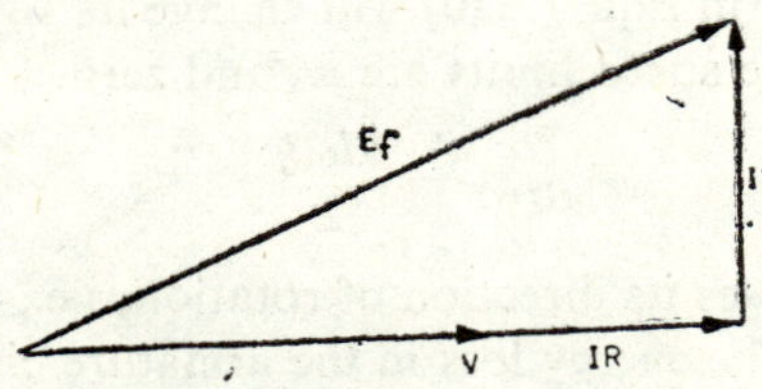

Fig. 6.19. Phasor diagram of voltages for Example 6-5.

From the phasor diagram shown in Fig. 6.19 E_f corresponding to the rated current at unity power factor

$$= \sqrt{(1330 + 751.88 \cdot 0.2)^2 + (751.88 \cdot 2)^2}$$

$$= 2110.17 \text{ V}$$

$$I_{br} = \frac{\omega L_{af} I_f}{\sqrt{2}\sqrt{r^2 + (\omega L_s)^2}} = \frac{E_f}{\sqrt{r^2 + x_s^2}}$$

Let $r = 0.2 + R$, where R is the extra resistance/phase added in stator circuit.

Hence $\qquad 751.88 = \dfrac{2110.17}{\sqrt{(0.2 + R)^2 + (2)^2}}$

Solving $\qquad R = 1.77$ ohms

Initial braking torque $T_{br} = \dfrac{3 I_{br}^2 r}{\omega}$

$$= \frac{3 \cdot (751.88)^2 \cdot 1.97 \cdot 60}{2\pi \cdot 200}$$

$$= 159{,}523.95 \text{ N-m}$$

6.6 Energy Relations During Braking

As during starting of motors, certain expressions can be obtained for energy lost by the motors during braking.

6.6.1 *DC Motors with Separate or Shunt Excitation*

Let us suppose that the braking takes place with no load on the motor.

The energy dissipated in the armature circuit during rheostatic braking is determined using Eqn. (5.20). It must be noted that during rheostatic braking $V = 0$ and the braking takes place from speed ω_0 to standstill.

Hence, $$W_{br(rhe)} = \int_{\omega 0}^{0} - J\omega \, d\omega$$

$$= \int_{0}^{\omega 0} \omega \, d\omega = \frac{J\omega_0^2}{2} \tag{6.17}$$

Again, it is seen that the loss in energy during rheostatic braking of a dc motor on no load will be equal to the kinetic energy possessed by the armature at the start of the braking process.

During reverse current braking, the applied voltage V is of the opposite polarity and hence ω_0 in Eqn. (5.20) will change its sign, while determining the loss in energy. The speed limits are ω_0 and zero.

$$W_{br(rev)} = \frac{3J\omega_0^2}{2} \tag{6.18}$$

If the motor reverses its direction of rotation, i.e., if the speed changes from $+\omega_0$ to $-\omega_0$, the energy loss in the armature circuit will be given by

$$W_{rev} = 4\frac{J\omega_0^2}{2} \tag{6.19}$$

6.6.2 Induction Motors

Let us assume that braking is performed with motor on no load.

During reverse current braking or plugging the slip changes from 2 (very nearly) to 1. Substituting these limits of slip in Eqn. (5.31), we get

$$W_{br(rev)} = \frac{J\omega_s^2}{2}(2^2 - 1^2) = \frac{3J\omega_s^2}{2} \tag{6.20}$$

The heat dissipated in the rotor circuit is, therefore, three times the kinetic energy at no load speed. Thus, this type of braking draws from the supply twice as much energy as there already exists in the drive and dissipates the whole lot in the rotor circuit resistances.

Incidentally, if the motor were reversed to run nearly at ω_s, the slip would vary from (nearly) 2 to (nearly) 0.

The loss during reversal of speed will then be given by

$$W_{rev} = \frac{J\omega_s^2}{2}(2^2 - 0^2) = 4\frac{J\omega_s^2}{2} \tag{6.21}$$

Both Eqn. (6.20) and Eqn. (6.21) represent the loss in the rotor circuit alone. The total energy dissipated by the motor during braking will be $\frac{3J\omega_s^2}{2}\left(1+\frac{R_1}{R_2}\right)$ joules and that during reversal of speed will be

$$\frac{4J\omega_s^2}{2}\left(1+\frac{R_1}{R_2}\right) \text{ joules.}$$

During rheostatic braking, the energy dissipated in the rotor circuit will be given by the stored kinetic energy available at the beginning of the braking process, i.e.,

$$W_{br(rhe)} = \frac{J\omega_s^2}{2} \tag{6.22}$$

Loss in energy in the stator, during rheostatic braking is given by,

$$W_{br(s)} = 3I_{ac\,(eq)}^2\, R_1 t_{br}, \tag{6.23}$$

where t_{br} denotes the braking time during which the motor speed falls from ω_s to zero.

Incidentally, the time during braking can be determined by substituting proper values of slip in Eqn. (5.13).

Example 6-6. A 230 V, 3-phase, 50 Hz, 4 pole, 10 A, 0.85 p.f. squirrel cage induction motor has a full load rated speed of 1440 rpm. The stator losses amount to 86.16 W. The total inertia of the motor together with the load is 0.0486 kg-m². Determine the number of starts by direct on line and stops by plugging per minute that this motor can make without exceeding the allowable temperature rise.

Solution:

$$\text{Input power} = \sqrt{3}\cdot 230\cdot 10\cdot 0.85$$
$$= 3386.16 \text{ W}$$

$$\text{Power transferred to rotor} = 3386.16 - 86.16$$
$$= 3300 \text{ W}$$

$$\text{Rotor copper loss} = \left(\frac{1500-1440}{1500}\right) 3300$$
$$= 132 \text{ W.}$$

Hence, permissible energy that can be dissipated in rotor in one minute, without exceeding the allowable temperature rise = 132.60 = 7920 J.

Total energy dissipated in rotor during starting and braking

$$= \tfrac{1}{2} J\omega_s^2 \,[1 - s^2 + (2-s)^2 - 1]$$
$$= 2\, J\omega_s^2\, (1-s)$$
$$= 2\cdot 0.0486\cdot\left(\frac{2\pi\cdot 50}{2}\right)^2\cdot 0.96$$
$$= 2302.38 \text{ J.}$$

Therefore, total number of starts and stops

$$= \frac{7920}{2302.38}$$
$$= 3.44 \text{ say, } 3.$$

Example 6-7. Determine the time required to bring to rest, by reverse current braking, a load of moment of inertia of 10 kg-m² from a speed of 950 rpm by means of a 440 V, 6 pole, 50 Hz, stator connected three phase induction motor having the following parameters:

$$R_1 = 0,\ R_2 = 0.2 \text{ ohm},\ X_1 = X_2 = 0.5 \text{ ohm}.$$

What resistance/phase must be added to the rotor circuit to bring the motor to rest in the quickest possible time and calculate this minimum time.

Solution:

$$T_{\max} = \frac{3V_1^2}{\omega_s \cdot 2(X_1 + X_2)}$$

$$= \frac{3\cdot(400/\sqrt{3})^2}{(2\pi\cdot 1000/60)\cdot 2\cdot 1}$$

$$= 763.94 \text{ N-m.}$$

The time required to stop the motor

$$t = \frac{J\omega_s}{2\,T_{\max}}\left[\frac{(2-s_1)^2-1}{2\,s_{\max T}} + s_{\max T}\log_e(2-s_1)\right]$$

Substituting, $J = 10$ kg-m², $s_1 = 0.05$ and $s_{\max T} = \dfrac{0.2}{1}$, we have,

$$t = \frac{10\cdot 2\pi\,(1000/60)}{2\cdot 763.94}\left[\frac{(1.95)^2-1}{2\cdot 0.2} + 0.2\log_e 1.95\right]$$

$$= 0.685\ (7.14) = 4.89 \text{ secs.}$$

$$(s_{\max T})_{\text{opt}} = \left[\frac{(2-s_1)^2-1}{2\log_e(2-s_1)}\right]^{1/2}$$

$$= \left(\frac{(1.95)^2-1}{2\log_e 1.95}\right)^{1/2}$$

$$= 1.45$$

$$(R_2)_{\text{opt}} = 1.45\cdot 1 = 1.45 \text{ ohms.}$$

Extra resistance to be added $= 1.45 - 0.2$

$= 1.25$ ohms.

Minimum time to bring the motor to rest

$$= \frac{10\cdot 2\pi\,(1000/60)}{2\cdot 763.94}\left[\frac{(1.95)^2-1}{2\cdot 1.45} + 1.45\log_e 1.95\right]$$

$$= 0.685\ (1.935)$$

$$= 1.325 \text{ seconds.}$$

6.6.3 *Synchronous Motors*

Just as in the case of induction motors the energy dissipated in the rotor circuit during rheostatic braking will be given by

$$W_{\text{br}} = \frac{J\omega_s^2}{2} \qquad (6.24)$$

where ω_s represents the synchronous speed.

Loss in energy in the stator, during braking will be expressed as:

$$W_{\text{br(s)}} = 3\,I_{\text{br}}^2\,r t_{\text{br}}, \qquad (6.25)$$

and I_{br} for a particular excitation current will remain constant.

6.7 Dynamics of Braking

The torques acting during the braking period can, in general, be classified as follows:

(i) Motor developed torque $-T(\omega)$, the negative sign denoting that it is a braking torque.

(ii) Torque due to friction or friction brake T_F, which is normally assumed to be a constant.

(iii) Load torque T_L, which if it has some active torque should be written in the form $T_{LP}(\omega) \pm T_{LA}(\omega)$ so as to indicate the passive and active portions of the load torque separately. The choice of sign to be used with the active load torque will depend upon the direction in which the unbalanced load is acting. If an unbalanced load is being lifted, the load will add to the total braking effort, and, hence, the sign will be positive.

(iv) Inertia torque $J \dfrac{d\omega}{dt}$.

Let us consider an example in which the motor is to be stopped by an electromagnetic torque developed by it, which is directly proportional to the speed and by some form of frictional brake. The load torque on the motor is assumed to be independent of speed. The torque equation then becomes,

$$-T(\omega) = T_L + T_F + J\frac{d\omega}{dt} \tag{6.26}$$

i.e.,

$$-K\omega = T_L + T_F + J\frac{d\omega}{dt}, \text{ where } K \text{ is a constant}$$

i.e.,

$$dt = -J\frac{d\omega}{K_\omega + T_L + T_F}.$$

Therefore, the time required to brake the motor from an initial speed ω_1 to final speed ω_2 in the same direction will be given by

$$t = -J\int_{\omega_1}^{\omega_2} \frac{d\omega}{K\omega + T_L + T_F}$$

$$= \frac{J}{K}\log_e\left(\frac{K\omega_1 + T_L + T_F}{K\omega_2 + T_L + T_F}\right) \tag{6.27}$$

Hence, the time required to bring the motor to rest is obtained by substituting $\omega_2 = 0$ in Eqn. (6.27).

$$t_r = \frac{J}{K}\log_e\left(\frac{K\omega_1 + T_L + T_F}{T_L + T_F}\right) \tag{6.28}$$

The number of revolutions made by the motor to come to standstill, may be calculated as follows:

$$N = \frac{1}{2\pi}\int_0^t \omega_2\, dt, \tag{6.29}$$

where, N is the number of revolutions made during braking from speed ω_1 to speed ω_2.

From Eqn. (6.27), we get

$$\omega_2 = \frac{1}{K}\left[(K\omega_1 + T_L + T_F)\, e^{-Kt/J} - (T_L + T_F)\right] \tag{6.30}$$

Substituting Eqn. (6.30) in Eqn. (6.29), we get

$$N = \frac{1}{2\pi K}\left[\frac{J}{K}(K\omega_1 + T_L + T_F)(1 - e^{-Kt/J}) - (T_L + T_F)\,t\right] \tag{6.31}$$

Replacing t by t_r, we obtain the number of revolutions made to come to rest as

$$N_r = \frac{1}{2\pi K}\left[\frac{J}{K}(K\omega_1 + T_L + T_F)(1 - e^{-Kt_r/J}) - (T_L + T_F)\,t_r\right] \tag{6.32}$$

Example 6-8. A 6-pole, 50 Hz synchronous motor coupled to a load has a moment of inertia of 540 kg-m². If the load torque is independent of speed and the frictional torque is 300 N-m, calculate the time taken by the motor to come to stop and the number of revolutions made during:

(a) rheostatic braking, the field current and braking resistance being kept constant at values which give initial electrical braking torque of 6000 N-m,

(b) plugging which produces a constant braking torque of 3000 N-m.

Solution: (a) Initial braking torque $K\omega = 6000$ N-m

$$\omega = \frac{2\pi\cdot 50}{3} \text{ rad/sec.}$$

Hence,

$$K = \frac{6000\cdot 3}{2\pi\cdot 50} = 57.3$$

$$t_r = \frac{J}{K}\log_e\left(\frac{K\omega + T_F}{T_F}\right)$$

$$= \frac{540}{57.3}\log_e\frac{6300}{300} = 28.69 \text{ sec.}$$

$$N_r = \frac{1}{2\pi K}\left[\frac{J}{K}(K\omega + T_F)(1 - e^{-Kt_r/J}) - T_F\cdot t_r\right]$$

$$= \frac{1}{2\pi\cdot 57.3}\left[\frac{540}{57.3}\cdot 6300\,(1 - e^{-57.3\cdot 28.69/540}) - 300\cdot 28.69\right]$$

$$= 133.15$$

(b) Total braking torque = 3300 + 300 = 3600 N-m. Since the braking torque is constant, it will produce a uniform retardation of β rad/sec².

$$\beta = \frac{T_B}{J} = \frac{3600}{540} = 6.67 \text{ rad/sec}^2$$

$$\omega = \frac{2\pi\cdot 50}{3} \text{ rad/sec.}$$

Hence, time taken by the motor to come to rest

$$= \frac{2\pi\cdot 50}{3\cdot 6.67} = 15.71 \text{ sec.}$$

Number of revolutions made before coming to stop

$$= \text{(Average speed in rev/sec)} \cdot \text{(time taken to stop in sec)}$$

$$= \frac{1000}{60} \cdot \frac{1}{2} \cdot 15.71$$

$$= 130.9.$$

PROBLEMS

1. Rheostatic braking is emlpoyed with a 4-pole dc separately excited motor driving a load. The armature is wave wound and has 251 conductors. The moment of inertia of the motor as well as the load is 350 kg-m^2. The resistance of the armature circuit winding is 0.02 ohm. Assuming that the flux per pole is constant at 0.02 Wb, calculate the time taken to reduce the speed from 100 rpm to 1 rpm. The maximum permissible armature current is 300 A and may be assumed to be constant until the entire external armature circuit resistance is cut out.
2. A dc shunt motor has a magnetization curve at 1000 rpm as given below:

Field current (A)	2.5	5.0	7.5	10	15	20	25
EMF (V)	40	80	117	142	168	184	215

   The armature resistance is 0.05 ohm and the field resistance is 110 ohms. The speed of the motor is reduced from 1000 to 540 rpm by reverse current braking. The supply voltage is 220 V and the armature current is not to exceed 202 A. The armature circuit resistance remains constant during braking. Calculate the time taken when the moment of inertia of the armature together with the load is 80 kg-m^2.
3. A 220 V dc shunt motor having an efficiency of 88 per cent drives a hoist having an efficiency of 75 per cent. Calculate the current drawn from the supply to raise a load of 400 kg at 2.5 m per second. What resistance must be added to the armature circuit in order to lower the load at 2.5 m per second, using rheostatic braking. Assume that the efficiency of the hoist and the dc machine remain the same as before.
4. A 50 kW, 400 V, 3-phase, 4-pole, 50 Hz induction motor has a full load slip of 4 per cent. If the ratio of standstill reactance to resistance per phase of rotor is 5, calculate the torque developed at the start of reverse current braking at rated speed. Neglect stator impedance and magnetizing current.
5. A 440 V, star connected, 3-phase, 6-pole, 50 Hz induction motor having the following equivalent circuit parameters in ohms per phase referred to the stator is used for regenerative braking: $R_1 = 0.1$; $R_2 = 0.08$; $X_1 = X_2 = 0.2$, $X_m = 10$. Determine:
   (a) the maximum overhauling torque which it can hold, and
   (b) the speed at which it will hold a load with a torque corresponding to 50 per cent of what has been determined in (a).
6. A 3-phase, star connected 440 V squirrel cage induction motor has the following equivalent circuit parameters in ohms per phase referred to stator: $R_1 = 0.1 = R_2$; $X_1 = 0.4 = X_2$; $X_m = 10$. Determine:
   (a) the starting current of the motor, when switched direct on line, and
   (b) the stator current a t the start of reverse current braking.

7. A 3-phase, star connected, 400 V, 50Hz, 4-pole induction motor has the following equivalent circuit parameters referred to stator in ohms per phase: $R_1 = 0.8$, $R_2=0.3$, $X_1=X_2=2$; $X_m=48$. An external resistance of 2 ohms per phase referred to stator has been inserted in the rotor circuit in order to brake the motor at 1440 rpm by means of dc rheostatic braking. Determine:

   (a) the initial braking torque, and
   (b) the magnitudes of dc currents to be fed to the stator for different connections of the stator winding.

8. A 440 V, 3-phase, 50 Hz, 6-pole induction motor has the following equivalent circuit constants per phase referred to stator in ohms: $X_1=X_2=1.5$; $X_m=45$. Stator resistance is negligible. Determine the magnitude of the rotor circuit resistance which enables the motor together with the load to be brought to rest by reverse current braking in the quickest possible time from a speed of 950 rpm. The total inertia of the motor and the load may be taken as 5 kg-m^2.
   Calculate the minimum time required for braking the motor to stop.

9. A motor driving a load is stopped by a mechanical brake which provides a constant braking torque equal to twice the rated torque of the motor. Determine the time taken by the motor to stop and the number of revolutions made before coming to rest from a speed of N rpm, if the load torques on the motor are given by (a) $T_L=0$ and (b) $T_L=1.5$ rated torque.

10. A motor having an inertia of 1.2 kg-m^2 and rotating at 960 rpm drives a hoist at 1 m per sec. The maximum load on the cables is 2000 kg and there is no counterweight. A constant braking torque of twice the magnitude of the above maximum load torque is being provided by means of a braking device. The friction torque of the hoist drive is 25 per cent of the load torque. Calculate the time taken to stop and the number of revolutions made by the motor to stop when

    (a) the load is moving upwards, and
    (b) the load is moving downwards.

Answers

1. 14.71 secs
2. 64.1 secs
3. 67.56 A, 7.47 ohms
4. 174.12 Nm
5. (a) 2844 Nm (b) 1060 rpm
6. (a) 93 Nm
7. 3.58 ohms, 1.74 secs

Chapter 7

Rating and Heating of Motors

The size and rating of a motor to be used as a drive element in a particular application depend upon the following factors:

(i) Heating effects in the motors
(ii) Loading conditions, and classes of duty
(iii) Load inertia or inertias
(iv) Environmental conditions.

7.1 Heating Effects

The heating of a machine is a function of the losses within it that are developed as heat. The cooling depends upon the facilities for heat dissipation, i.e, ventilation, to outside media such as air, oil or solids. The temperature rise depends on the inter-relationship between heating and cooling. Under steady state conditions, the final temperature rise is reached when the rates of production and dissipation of heat are equal. Electrical machines are designed for a limited temperature rise. In fact, the continuous rating of a machine is that rating for which the final temperature rise is equal to or just below the permissible value in temperature rise for the insulating material used in protecting the conductors. When the machine is overloaded for such a long time that its final temperature rise is higher than the allowable limit, it is likely to be damaged. In worst cases, this will result in an immediate thermal breakdown of the insulating material which will cause a short circuit in the motor, thus putting a stop to its functioning. The short circuit may also lead to a fire. In less severe cases immediate thermal breakdown of the insulating material may not occur, but the quality of the insulation will deteriorate such that thermal breakdown with future overloads or even normal loads might soon occur, thus shortening the useful life of the machine.

For the permissible temperature limits on commonly used insulating materials, ISS: 1271 should be consulted. Since class A insulated machines are not considered as standard machines, the insulation-system classes of main interest for machines used in industry are class E, class B, class F and class H. ISS: 4722 gives the limits of permissible temperature rise for the various parts of rotating electrical machines. Fairly detailed distinctions are made as regards the type of machine, machine part involved, method of temperature, measurement, the type of cooling (air cooled, hydrogen cooled etc.) and class of duty.

7.1.1 *Heating and Cooling Curves*

In so far as a machine can be considered as a homogeneous body developing heat internally at uniform rate and dissipating heat proportionately to its temperature rise, the relationship between temperature rise and time can be shown to be an exponential function.

Let p = heat developed, joules/sec or watts;
G = weight of active parts of machine, kg;
h = specific heat, J per kg per °C;
S = cooling surface, m²;
λ = specific heat dissipation or emissivity, J per sec. per m² of surface per °C difference between surface and ambient cooling medium.
θ = temperature rise, °C;
θ_m = final steady temperature rise, °C;
t = time, sec.;
τ = heating time constant, sec.;
τ' = cooling time constant, sec.;

Assume that a machine attains a temperature rise θ after the lapse of time t sec. In an element of time dt a small temperature rise $d\theta$ takes place; the heat developed is $p \cdot dt$, the heat stored is $Gh \cdot d\theta$, and the heat dissipated in $S\theta\lambda \cdot dt$. Since the heat stored and dissipated add up to the total heat developed.

$$Gh \cdot d\theta + S\theta\lambda \cdot dt = p \cdot dt$$

or

$$\frac{d\theta}{dt} + \theta \cdot \frac{S\lambda}{Gh} = \frac{p}{Gh}$$

The solution of this differential equation is

$$\theta = \frac{p}{S\lambda} + Ke^{-(S\lambda/Gh)t}$$

where K is a constant of integration determined by initial conditions.

Let us suppose that the initial temperature rise to be zero at $t = 0$. Then $0 = \frac{p}{S\lambda} + K$, so that $K = -\frac{p}{S\lambda}$.

Hence

$$\theta = \frac{p}{S\lambda}\left(1 - e^{-(S\lambda/Gh)t}\right) \tag{7.1}$$

When $t = \infty$, $\theta = \frac{p}{S\lambda} = \theta_m$, the final steady temperature rise. Denoting, therefore,

$$\frac{p}{S\lambda} = \theta_m \quad \text{and} \quad Gh/S\lambda = \tau \tag{7.2}$$

we have, $$\theta = \theta_m(1 - e^{-t/\tau}) \quad (7.3)$$

τ has the dimensions of time, and is called the heating time constant.

When $t = \tau$, the exponential term reduces to $1/e = 0.368$, so that $\theta = 0.632\theta_m$, i.e. after a time τ from the start, θ attains 63.2 per cent of its final steady value. The temperature rise vs time curve is shown in Fig. 7.1.

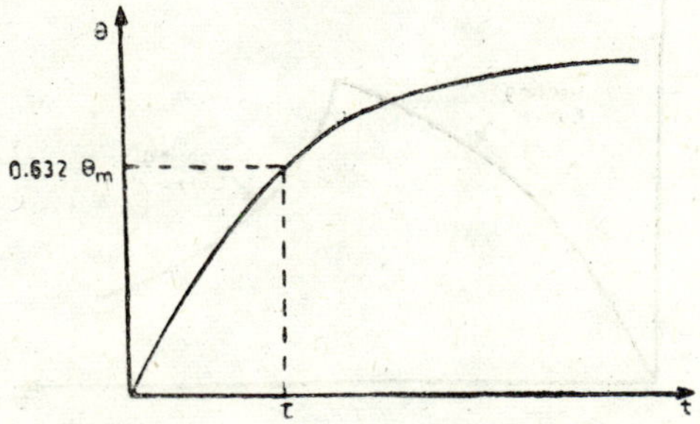

Fig. 7.1. Temperature rise-time curve.

Good ventilation gives a small time constant, since τ is inversely proportional to λ, which is large. Thus, for open machines of 10-20 kW rating τ may be of the order of 25 minutes and for machines of medium power rating (500-800 kW) it may be 2-2.5 hours. With large or totally enclosed machines, the heating time constant may reach several hours or even days.

When a hot body is cooling due to a reduction or cessation of the losses developed in it, the temperature time curve is again an exponential function, viz.

$$\theta = \theta_f + (\theta_m - \theta_f)\, e^{-t/\tau'} \quad (7.4)$$

where, θ_f the final temperature drop (the temperature at which whatever heat is generated is dissipated) given by $\frac{p}{S\lambda'}$, in which λ' is the rate of heat dissipation while cooling;

θ_m—the temperature rise above ambient in the body at time $t = 0$ and

τ'—cooling time constant given by $\frac{Gh}{S\lambda'}$.

The ventilation in a machine which is cooling at rest will be inferior to that when running. Hence τ', the cooling time constant, will normally differ from τ.

If motor were disconnected from the supply during cooling, there would be no losses taking place and hence, final temperature reached will be the ambient temperature.

Therefore $\theta_f = 0$ and Eqn (7.4) becomes

$$\theta = \theta_m e^{-t/\tau'} \quad (7.5)$$

At $$t = \tau' \qquad \theta = 0.368\,\theta_m \quad (7.6)$$

Cooling time constant is, therefore, defined as the time required to cool the machine down to 0.368 times the initial temperature rise above ambient temperature.

Heating and cooling time curves are shown in Fig. 7.2.

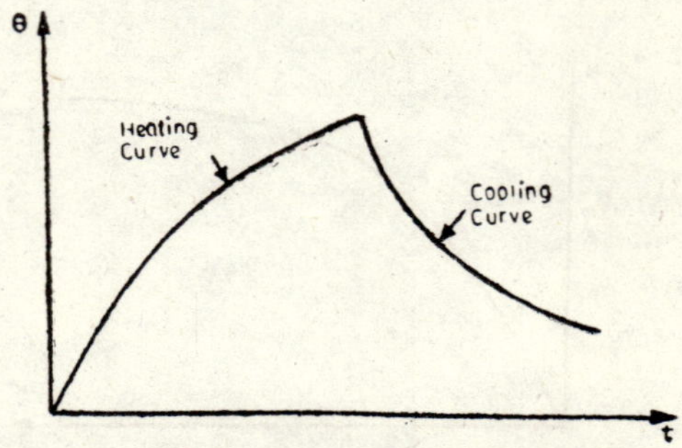

Fig. 7.2. Heating and cooling time curves.

Example 7-1. The enclosure of a 10kW motor is equivalent to a cylinder of 70 cm diameter and 100 cm length. The motor weighs 500 kg. Assuming that the specific heat is 700 J/kg/°C and that the peripheral surface of the enclosure of the motor alone is capable of heat dissipation of 12.5 W/sq. m/°C, calculate the heating time constant of the motor and its final temperature rise. Efficiency of motor is 90 per cent.

Solution:

$$\text{Heating time constant } \tau = \frac{Gh}{S\lambda} \text{ sec.}$$

$$= \frac{500 \cdot 700}{\pi \cdot 0.7 \cdot 1 \cdot 12.5}$$

$$= 12732.4 \text{ sec.}$$

$$= 3.54 \text{ hr.}$$

$$\text{Losses in the motor} = \frac{10 \cdot 1000}{0.9} - 10000$$

$$= 1111 \text{ watts}$$

$$\text{Final temperature rise} = \frac{W}{S\lambda} = \frac{1111}{\pi \cdot 0.7.1 \cdot 12.5}$$

$$= 40.4°\text{C}.$$

7.2 Loading Conditions and Classes of Duty

An electric motor, under steady state operation, develops electromagnetic torque of such a magnitude which can counterbalance the actual load torque T_L of the connected equipment and an opposing torque T_{mech}, corresponding to the losses that take place in gear and transmission mechanisms. Under transient conditions, the motor torque has to overcome the inertia torque T_{dyn} also. Hence, in general, the torque developed by the motor should be expressed as:

$$T = T_L + T_{mech} + T_{dyn} \tag{7.7}$$

The combined load torque ($T_L + T_{mech}$) is determined from the torque-time plot of the connected load. A typical example is shown in Fig. 7.3(a). In order to determine the variation of inertia torque with respect to time, the speed-time curve, an example of which is shown in Fig. 7.3(b) and the moment of inertia of the rotating masses J must be known. Now, with the help of Eqn. (7.7) it will be possible to obtain the torque-time curve of the driving motor [Fig. 7.3(c)], which is called the duty cycle of the motor.

Since, initially J remains unknown, the torque-time graph can be plotted by taking into account only T_L and T_{mech}. After determining the power rating of the machine using the above torque-time graph, it can be corrected for the presence of inertia torque by increasing the rating obtained by 15-20 per cent. Knowing J of the chosen motor, the exact torque-time curve can

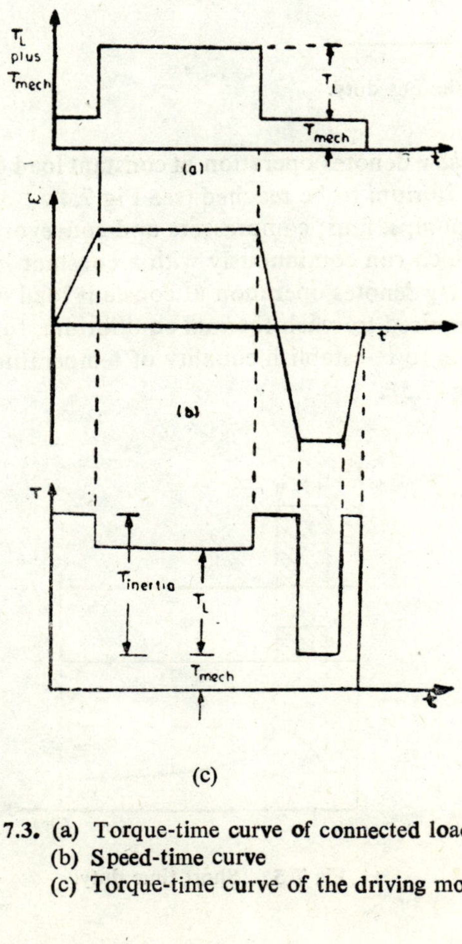

Fig. 7.3. (a) Torque-time curve of connected load
(b) Speed-time curve
(c) Torque-time curve of the driving motor.

be plotted and a more correct estimation of the rating of the driving motor can be made.

The rating of a motor selected from the viewpoint of heating depends on the load conditions or duty to which it is subjected. According to ISS: 4722, these operating conditions are classified into eight classes of duty, depending on the duration and nature of the load, viz., continuous duty, short time duty, intermittent periodic duty, intermittent periodic duty with starting, intermittent periodic duty with starting and braking, continuous duty with intermittent periodic loading, continuous duty with starting and braking and continuous duty with periodic speed changes.

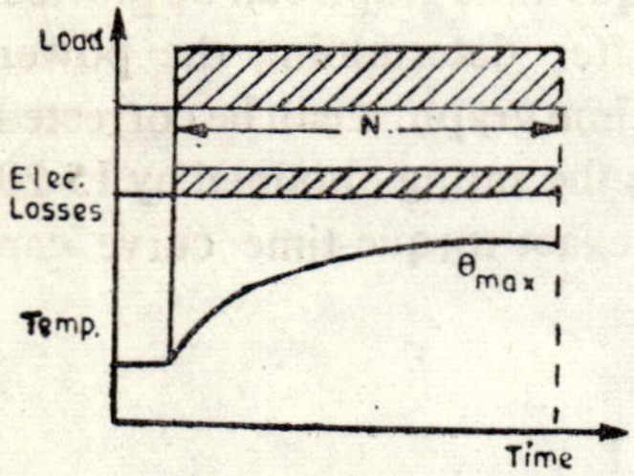

Fig. 7.4. Continuous duty.

N indicates the duration for which operation under rated conditions takes place and θ_{max}, the maximum temperature attained during the duty cycle.

Continuous duty denotes operation at constant load of sufficient duration for thermal equilibrium to be reached (see Fig 7.4).

Centrifugal pumps, fans, compressors and conveyors are some examples of equipment which run continuously with a constant load.

Short time duty denotes operation at constant load during a given time, less than that required to reach thermal equilibrium, followed by a rest of sufficient duration to re-establish equality of temperature with the cooling medium (see Fig. 7.5).

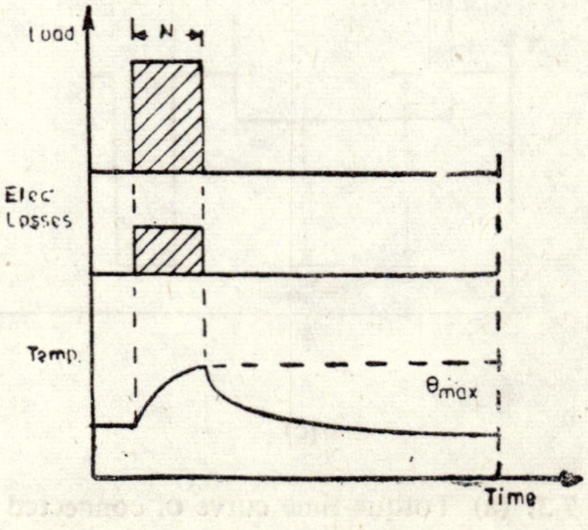

Fig 7.5. Short time duty.

The recommended values for the short time duty are 10, 30, 60 and 90 minutes.

Motors used for opening and closing weirs, lockgates and bridges, motors employed in battery-charging units etc. are rated for such a duty.

Intermittent periodic duty denotes a sequence of identical duty cycles, each consisting of a period of operation at constant load and a rest period, these periods being too short to attain thermal equilibrium during one duty cycle (see Fig. 7.6). The starting current does not significantly affect the temperature rise for this type of duty. Unless otherwise specified, the duration of the duty cycle is 10 minutes. The recommended values for the cyclic duration factor are 15, 25, 40 and 60 per cent.

R indicates the duration of rest, the period in which there is a total absence of all motion and of all electrical or mechanical supply. Cyclic duration factor $= \frac{N}{N+R}$.

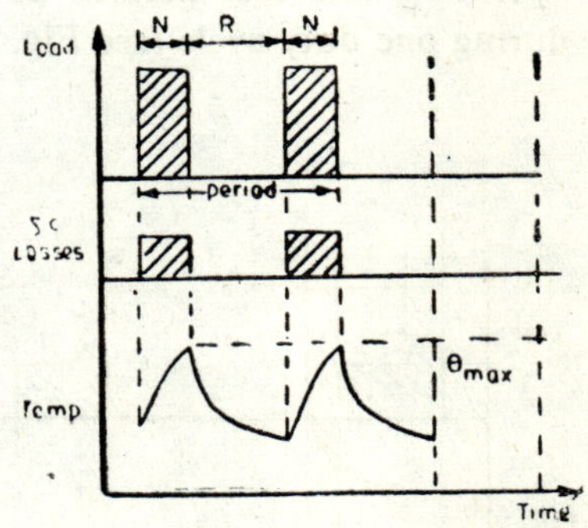

Fig. 7.6, Intermittent periodic duty.

Motors that are used in different kinds of hoisting mechanisms and those used in trams, trolley buses etc. are subjected to intermittent duty.

Intermittent periodic duty with starting indicates a sequence of identical duty cycles each consisting of a period of starting, a period of operation at constant load and a rest period, the operating and rest periods being too short to attain thermal equilibrium during one duty cycle (see Fig. 7.7).

D denotes the duration of starting. Cyclic duration factor $= \frac{D+N}{D+N+R}$.

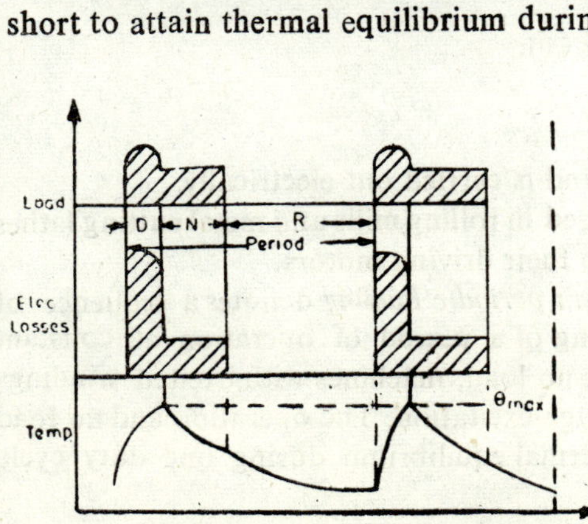

Fig. 7.7. Intermittent periodic duty with starting.

In this type of duty the stopping of the motor is obtained by natural deceleration after disconnecting the electrical supply by means of mechanical brake which does not cause additional heating of the windings.

This type of duty is characterized by the cyclic duration factor, the number of duty cycles per hour (c/h) and factor of inertia (FI), which is the ratio of the combined inertia of the motor and load to the motor inertia.

Motors that drive metal cutting lathes and certain auxiliary equipment of rolling mills are subjected to such operating conditions.

Intermittent periodic duty with starting and braking denotes a sequence of identical duty cycles each consisting of a period of starting, a period of operation at constant load, a period of braking and a rest period. The operating and rest periods are too short to obtain thermal equilibrium during one duty cycle (see Fig. 7.8).

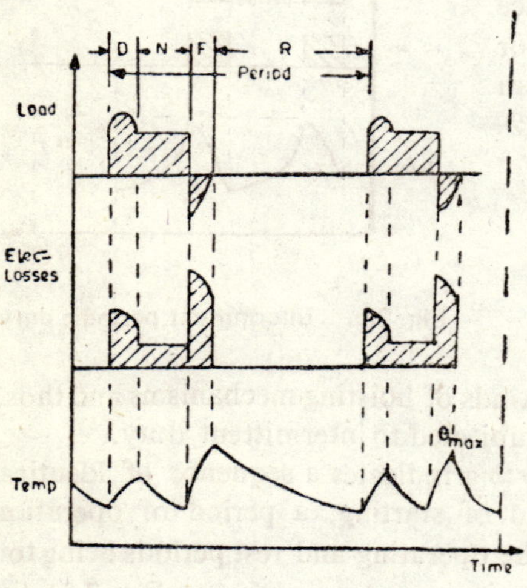

Fig. 7.8. Intermittent periodic duty with starting and braking.

F indicates duration of braking.
Cyclic duration factor

$$= \frac{D+N+F}{D+N+F+R}.$$

In this duty braking is rapid and is carried out electrically.

Certain auxiliary equipment used in rolling mills and metal cutting lathes offer such operating conditions to their driving motors.

Continuous duty with intermittent periodic loading denotes a sequence of identical duty cycles each consisting of a period of operation at constant load and a period of operation at no load, machines with excited windings having normal no load rated voltage excitation. The operation and no load periods are too short to attain thermal equilibrium during one duty cycle (see Fig. 7.9).

Unless otherwise specified the duration of the duty cycle is 10 minutes.

V = Duration of no load operation.

Cyclic duration factor $= \frac{N}{N+V}$.

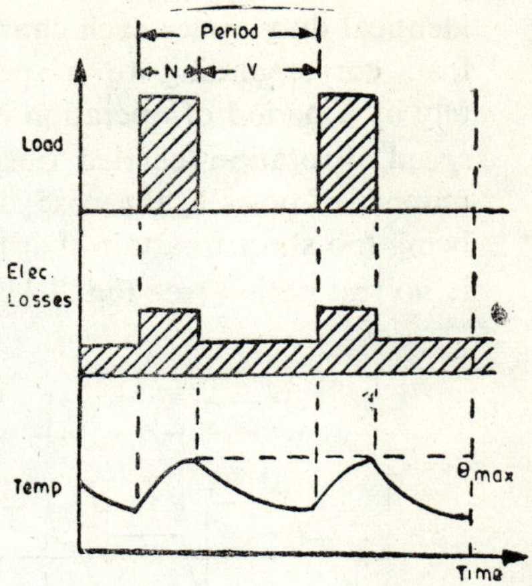

Fig. 7.9. Continuous duty with intermittent periodic loading.

The recommended values of cyclic duration factor are 15, 25, 40 and 60 per cent.

This type of duty is distinguished from intermittent periodic duty by the fact that after a period of operation at constant load follows a period of no load operation instead of rest.

Continuous duty with starting and braking denotes a sequence of identical duty cycles each consisting of a period of starting, a period of operation at constant load and a period of electrical braking (see Fig. 7.10). There is no period of rest.

This type of duty is also indicated by the number of cycles per hour and the factor of inertia.

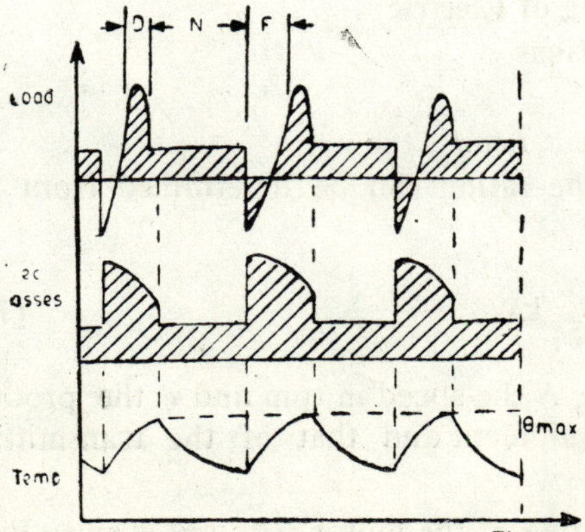

Cyclic duration factor = 1.

Fig. 7.10. Continuous duty with starting and braking.

Continuous duty with periodic speed changes indicates a sequence of identical duty cycles each consisting of a period of operation at constant load corresponding to a specified speed of rotation, followed immediately by a period of operation at another load corresponding to a different speed of rotation (carried out, for example, by means of change of the number of poles in the case of induction motors), the operating periods being too short to attain thermal equilibrium during one duty cycle. There is no rest period (see Fig. 7.11).

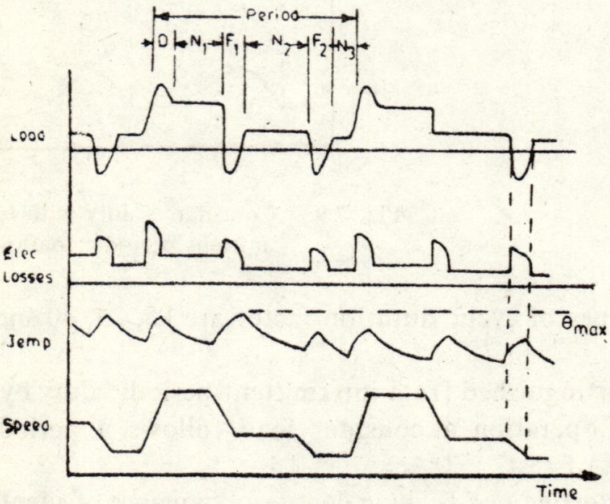

F_1, F_2 — Electric braking.
N_1, N_2, N_3 = Operation under rated conditions.
Cyclic duration factors

$$= \frac{D+N_1}{X}$$

$$= \frac{F_1+N_2}{X} = \frac{F_2+N_3}{X}.$$

where

$$X = D+N_1+F_1+N_2+F_2+N_3.$$

Fig. 7.11. Continuous duty with periodic speed changes.

For this duty type the number of duty cycles per hour and the factor of inertia together with the load have to be indicated. In addition, the cyclic duration factor should be indicated for each speed.

7.3 Determination of Power Rating of Electric Motors for Different Applications

7.3.1 Continuous Duty and Constant Load

For most of the applications, the rating can be determined from the expression

$$P = \frac{TN}{975\eta} \text{ kW}, \tag{7.8}$$

where T is the load torque in kg-m, N the speed in rpm and η the product of the efficiency of the driven equipment and that of the transmitting device.

In case of linear motion, the rating of the motor required is given by

$$P = \frac{Fv}{102\eta} \text{ kW}, \tag{7.9}$$

where F is the force caused by the load in kg and v the velocity of motion of the load in m/sec.

The above expression is directly applicable in case of hoisting mechanisms. It is also suitable for lifts or elevators; but since a counterweight which balances the weight of the cage or car as well as one-half of the useful load will be always present, it should be modified as follows:

$$P = \frac{Fv}{2 \cdot 102\eta} \text{ kW}, \tag{7.10}$$

The velocity of the normal passenger lift cabins vary from 0.5 to 1.5 m/sec.

In case of pumps, the rating can be determined from

$$P = \frac{\rho HQ}{102\eta} \text{ kW}, \tag{7.11}$$

where ρ—density of the liquid pumped, kg/m^3,

H—gross head (static head + friction head), m

Q—delivery of the pump, m^3/sec.

η varies from 0.8 to 0.9 for reciprocating pumps and from 0.4 to 0.8 for centrifugal pumps.

Similarly, the rating of a fan motor is given by

$$P = \frac{Qh}{102\eta} \text{ kW}, \tag{7.12}$$

where Q is the volume of air or any other gas in m^3/sec and h is the pressure in mm of water or in kg/m^2. For small power fans, the efficiency η may be taken as 0.6 and for large power ones it may reach a value upto 0.8.

The rating of a motor used in metal shearing lathes can be found from

$$P = \frac{F \cdot v}{102 \cdot 60 \cdot \eta} \text{ kW}, \tag{7.13}$$

where Q—shearing force, kg;

v—velocity of shearing, m/min;

η—mechanical efficiency of the lathe.

7.3.2 Continuous Duty and Variable Load

The rating of a motor for such operating conditions can be determined on the basis of average losses. In this method it is assumed that the temperature rise attained by the motor with variable loading conditions over a certain period of time will be the same as that reached by the motor with a certain constant magnitude of load. The above will be true provided that the average losses in the motor for the same time are the same for both the conditions of operation.

Let us first assume a load-time graph as shown in Fig. 7.12. It has different load torques corresponding to different intervals of time, including a period of rest (motor de-energized from the supply).

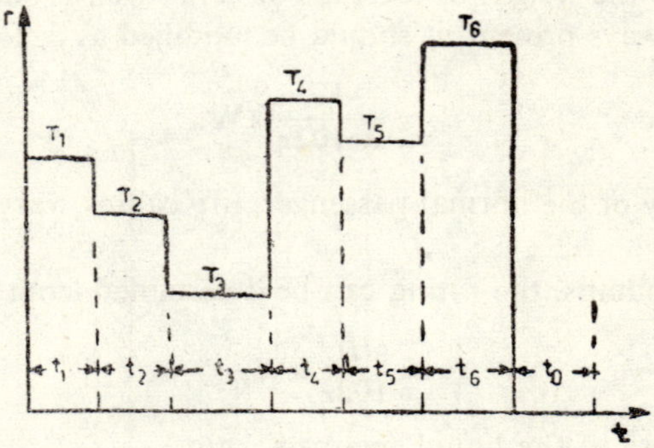

Fig. 7.12. Load torque vs. time.

Average losses over the total period of time

$$= \frac{W_{L_1} \cdot t_1 + W_{L_2} \cdot t_2 + \dots + W_{Ln} \cdot t_n}{t_1 + t_2 + \dots + t_n + t_0} \tag{7.14}$$

where W_{L_1}, W_{L_2}, ..., W_{Ln} are the losses that occur in the motor during the intervals t_1, t_2, ..., t_n etc. Note that no losses occur during the time interval t_0.

For an assumed constant current I_{eq} which causes the same average losses over the period of time considered, average losses can be written as $(W_c + I_{eq}^2 R)$, where W_c and R denote the constant losses (core + mechanical losses) and the resistance of armature circuit of the motor.

Also, losses at time interval $t_i = W_c + RI_i^2$, where I_i represents the load current during t_i.

$$W_c + I_{eq}^2 R = \frac{R \sum_{i=1}^{n} I_i^2 t_i}{t_1 + t_2 + \dots + t_n + t_0} + \frac{W_c (t_1 + t_2 + \dots + t_n)}{(t_1 + t_2 + \dots + t_n + t_0)} \tag{7.15}$$

Let us denote $\left(\frac{t_1 + t_2 + \dots + t_n}{t_1 + t_2 + \dots + t_0}\right) = \epsilon.$

$$\text{Then, } I_{eq}^2 R = \frac{R \sum_{i=1}^{n} I_i^2 t_i}{t_1 + t_2 + \dots + t_n + t_0} - W_c (1 - \epsilon) \tag{7.16}$$

Suppose that $W_c = \alpha W_{cu}$, where α is a fraction and W_{cu} is the copper losses (I^2R losses) corresponding to rated current of the machine, i.e., $W_c = \alpha W_{cu}$, which nearly equals to $\alpha I_{eq}^2 R$.

Substituting this in Eqn. (7.16), we have

$$I_{eq}^2 R[1 + \alpha(1 - \epsilon)] = R \cdot \frac{\sum_{i=1}^{n} I_i^2 t_i}{t_1 + t_2 + \ldots + t_n + t_0}$$

Hence, $$I_{eq} = \sqrt{\frac{\sum_{i=1}^{n} I_i^2 t_i}{t_1 + t_2 + \ldots + t_n + t_0}} \cdot \frac{1}{\sqrt{1 + \alpha(1 - \epsilon)}} \tag{7.17}$$

If the load were to have a no-load operation (a torque T_0 corresponding to constant losses in the machine) instead of the rest period,

$$I'_{eq} = \sqrt{\frac{\sum_{i=1}^{n} I_i^2 t_i}{t_1 + t_2 + \ldots + t_n + t_0}} \tag{7.18}$$

Eqn. (7.18) could be obtained either by equating the average losses over the period or by substituting $\epsilon = 1$ in Eqn. (7.17).

In case, neither a rest period nor a no-load period exists in the cyclic variable load,

$$I''_{eq} = \sqrt{\frac{\sum_{i=1}^{n} I_i^2 t_i}{t_1 + t_2 + \ldots + t_n}} \tag{7.19}$$

The above expression could be obtained by substituting $\epsilon = 1$ and $t_0 = 0$ in Eqn. (7.17).

If the load-time curve were not having constant load operations at various intervals of time and were to vary with time in an arbitrary manner as shown in Fig. 7.13 the term $\sum_{i=1}^{n} I_i^2 t_i$ in Eqns. (7.15) and (7.16) would be replaced by $\int_0^{t_n} i^2\, dt$.

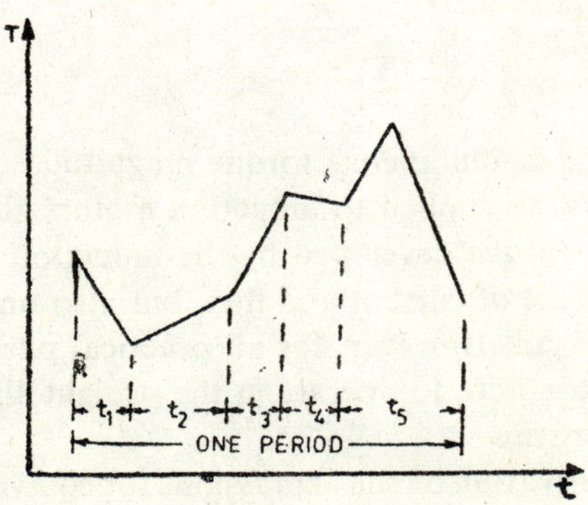

Fig.7.13. Varying load torque vs. time.

From Eqns. (7.17) and (7.18), we have

$$I'_{eq} : I_{eq} = \sqrt{1 + \alpha(1 - \epsilon)} : 1 \tag{7.20}$$

and from Eqns. (7.17) and (7 19), we get

$$I''_{eq} : I_{eq} = \sqrt{\frac{1 + \alpha(1 - \epsilon)}{\epsilon}} : 1 \tag{7.21}$$

Therefore, $I''_{eq} > I'_{eq} > I_{eq}$.

The equivalent currents, obtained from Eqns. (7.17) to (7.19) flowing continuously produce the same amount of heat in the motor as the actual load currents in the duty cycle.

The rated current selected for the motor should be equal to or greater than this equivalent current. It is also necessary to check the overload current capability λ_i of the motor.

$$\lambda_i \text{ must be} \geqslant \frac{I_{max}}{I_f}, \tag{7.22}$$

where I_{max} is the maximum current in the load curve and I_f, the rated current of the motor. λ_i for normal dc shunt motors varies from 2-2.5.

If Eqn. (7.22) were not satisfied, the rated current should be chosen on the basis of the overload current capability and not on the equivalent current.

The power rating P of the motor is found from the manufacturers' data corresponding to the chosen rated current and the fairly constant speed at which the load operates. This method of determining the power rating of the drive is called as the *method of equivalent current*.

In case of machines, whose flux remains constant irrespective of variations in load (dc shunt motors and separately excited motors of fixed excitation currents), the equivalent torque rating can be determined from any one of the above derived equivalent current equations.

$$T_{eq} = \sqrt{\frac{\sum_{i=1}^{n} T_i^2 t_i}{\sum_{i=1}^{n} t_i}}, \tag{7.23}$$

where T_i represents the instantaneous torque magnitudes.

Equation (7.23) can be applied to induction motors also, but with some approximations. The torque developed by the induction motor is not only dependent on the product of current and flux, but also on the power factor, which varies with the load. However, for all practical purposes, Eqn. (7.23) can be used if the motor were to operate in the straight line portion of its speed torque characteristics.

Since the entire derivation of the expressions for equivalent currents were on the assumption that the speed doesn't vary much with time (the constant losses of the machine were taken to be the same for various loads), Eqns.

(7.17), (7.18), and (7.19) are not valid for series motors, whose speed changes with load are quite significant.

For a fairly constant speed machine, the equivalent power rating can be obtained from the expression

$$P_{eq} = \frac{T_{eq}N}{975}, \tag{7.24}$$

where T_{eq} is in kg-m and N, the speed at which the load operates in rpm.

This method, in which the motor rating is determined from the load torque vs. time curve, is named as the *method of equivalent torque.*

Since, it is assumed that the speed at which the load operates is fairly constant, the power P will be directly proportional to torque T and, hence, substituting P for T in Eqn, (7.23), we get

$$P_{eq} = \sqrt{\frac{\sum_{i=1}^{n} P_i^2 t_i}{\sum_{i=1}^{n} t_i}} \tag{7.25}$$

This method of determining the rating of the motor from the load power vs. time cycle is termed as the *method of equivalent power.*

It is to be noted that the methods discussed above, for choosing the rating of the motor are all based on the assumption that the cooling conditions remain the same during the period of operation. This is not true; in fact, during starting, braking and stops, the cooling conditions differ considerably. In order to estimate the rating of motors used in such operations, suitable correction factors C_1 for starting and braking and C_2 for stops are introduced. For example, the equivalent torque of a motor supplying a load as shown in Fig. 7.14 is given by

$$T_{eq} = \sqrt{\frac{T_1^2 t_s + T_2^2 t_r + T_3^2 t_b + 0}{C_1 t_s + t_r + C_1 t_b + C_2 t_o}} \tag{7.26}$$

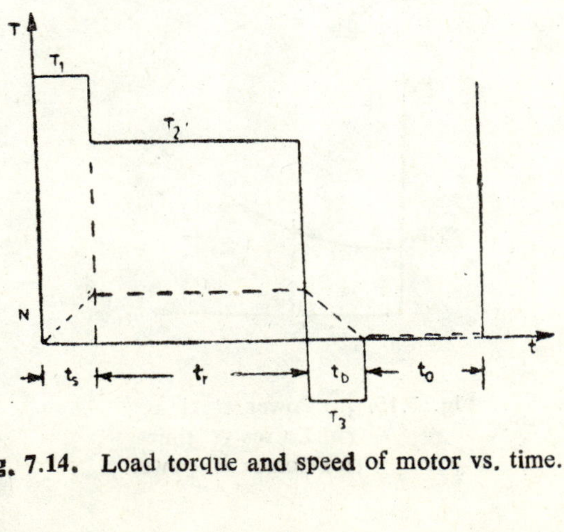

Fig. 7.14. Load torque and speed of motor vs. time.

However, no fixed values are used for C_1 and C_2 by manufacturers. The commonly used values are:

$C_1 = 0.5$ and $C_2 = 0.25$ in case of induction motors and
$C_1 = 0.75$ and $C_2 = 0.50$ for dc motors.

Yet another method to determine the rating of a motor subjected to a cyclic variable load is known as *average losses method.*

From the given load power vs. time curve (Fig. 7.15a), the average power P_{av} is first determined

$$P_{av} = \frac{\sum_{i=1}^{n} P_i t_i}{\sum_{i=1}^{n} t_i} \tag{7.27}$$

Then an approximate motor rating is chosen from the formula $P_r = kP_{av}$, where k varies from 1.1 to 1.3. The losses vs load power curve for the motor of rating P_r has to be plotted using either the efficiency vs. load curve of the motor or from design data. By constructing losses vs. time curve (Fig. 7.15b) corresponding to the loading cycle given, it is possible to get the average losses occurring over the period of the load cycle. From $W_L = f(P)$ curve (Fig. 7.15c) determine the value of P_{av} corresponding to the magnitude of average losses. This value of P_{av} should be as far as

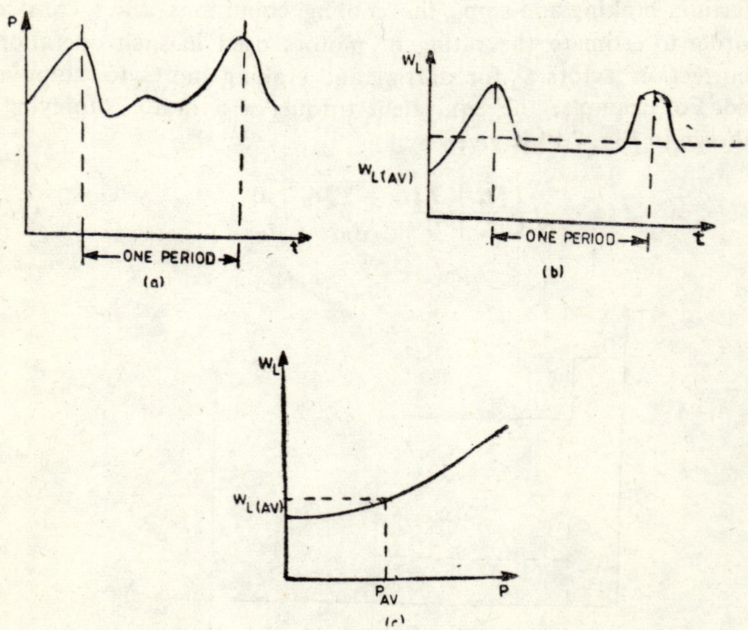

Fig. 7.15. (a) Power vs. time
(b) Losses vs. time
(c) Losses vs. power.

possible near to that of P_r. If it is $> P_r$, we have to choose the next higher rating motor available. The above method gives more accurate results, but requires additional data.

It is important to note that both the methods could be used only if the period of duty cycle were small compared to the heating time constant of the motor.

Example 7-2. The speed and torque demanded from a dc motor driving a load vary during the cycle as shown in Fig. 7.16(a) and (b) respectively. Determine the

(i) Peak kW rating of the motor,

(ii) Equivalent continuous rating of the motor based on rms torque, and

(iii) Equivalent continuous rating of the motor based on rms power.

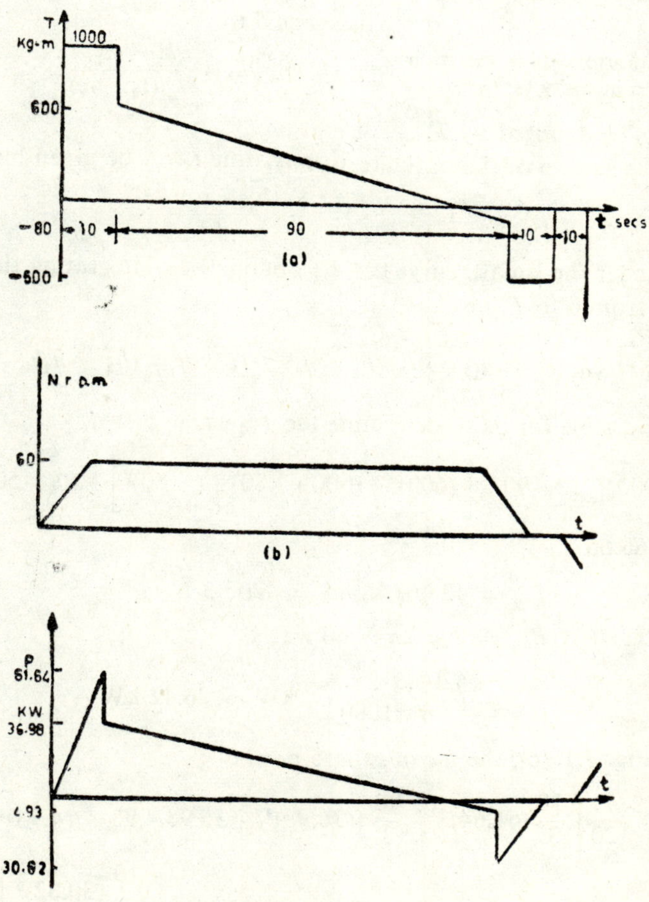

Fig. 7.16. (a) Torque-time variation
(b) Speed-time variation
(c) Power-time variation.

Solution: The power-time graph is plotted by multiplying the speed in rad/sec. and torque in N-m at every instant of time.

(i) The peak kW of the motor is found from the Fig. 7.16(c) as 61.64 kW.

(ii) $T_{eq} = \sqrt{\int_0^n T_i^2\, dt/(t_1 + t_2 + \ldots + t_n)}$

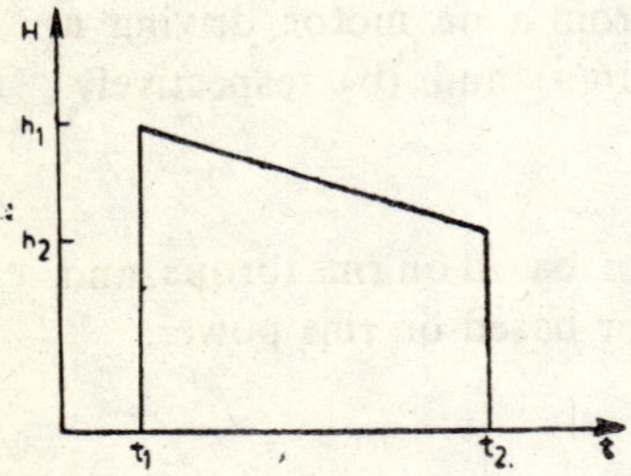

Fig. 7.17. Uniform sloping variation of a quantity vs. time.

Since, we have a period during which torque varies uniformly (straight sloping curve), it is necessary to find the area under the squared curve of a sloping straight line in order to determine the integral. It is done as follows:

Let h_1 and h_2 be the ordinates at time t_1 and t_2 (Fig. 7.17). At any time t between t_2 and t_1, the ordinate will have a magnitude equal to

$$h_1 + \frac{(h_2 - h_1)\, t}{(t_2 - t_1)}$$

Let $(t_2 - t_1)$ be denoted by T.

Then, the square of the ordinate at any time t will be given by

$$h_1^2 + \frac{2h_1\,(h_2 - h_1)\, t}{T} + \frac{(h_2 - h_1)^2\, t^2}{T^2}.$$

The area under the square curve can be obtained by integrating the above expression from 0 to T, i.e.

$$\int_0^T \{h_1^2 + 2h_1\,(h_2 - h_1)\, t/T + (h_2 - h_1)^2\, t^2/T^2\}\, dt = (h_1^2 + h_1 h_2 + h_2^2)\, T/3$$

Now, it is possible for us to determine the T_{eq} as,

$$T_{eq}^2 = \left\{ (1000)^2 \cdot 10 + 90/3 \left[(600)^2 + (600)(-80) + (-80)^2 \right] + 10(-500)^2 \right\} \frac{1}{120}$$

$$= 183766.66$$

$$T_{eq} = 428.68 \text{ kg-m} = 4205.3 \text{ N-m}$$

$$P_{eq} = T_{eq} \cdot 2\pi\, N/60 \text{ watts}$$

$$= \frac{4205.3 \cdot 2\pi \cdot 60}{60 \cdot 1000} \text{ kW} = 26.42 \text{ kW}$$

(iii) From Fig. 7.16(c), the mean square power

$$P_{eq}^2 = \frac{1}{120} \left[\frac{10}{3} (61.64)^2 + \frac{90}{3} \{(36.98)^2 - 36.98.4.93 + (4.93)^2\} + \frac{10}{3} (30.82)^2 \right]$$

$$= 434.31$$

$$P_{eq} = 20.8 \text{ kW}$$

7.3.3 Rating for Other Duty Cycles

An electric motor of rated power P_r subjected to its rated load continuously gets heated upto its permissible temperature rise. If such a machine is required to drive lighter loads, such as short time duty or intermittent duty loads, obviously, it can be loaded to a greater power P_x. Let us try to determine the values of P_x corresponding to the various duty cycles explained earlier.

(i) *Short time rating*: Let us assume that the heating of the motor is proportional to losses W_L, the duration of the short time load P_x is N and the heating time constant of the motor be τ. The temperature rise (θ'_m) attained by the motor should not exceed that reached when loaded to its continuous rating P_r. θ'_m is the permissible final temperature rise of the motor, which, in fact, is never reached, but serves as a measure of the losses and the permissible short time rating P_x of duration N. Since the temperature rise follows Eqn. (7.3), we have

$$\theta_m = \theta'_m (1 - e^{-N/\tau}) \tag{7.28}$$

or

$$\frac{\theta'_m}{\theta_m} = \frac{W_{Lx}}{W_{Lr}} = \frac{1}{(1 - e^{-N/\tau})}, \tag{7.29}$$

where W_{Lx} and W_{Lr} are the losses corresponding to the ratings P_x and P_r.

Suppose that the losses at rated load W_{Lr} is written as

$$\begin{aligned} W_{Lr} &= W_c + W_{cu} \\ &= \alpha W_{cu} + W_{cu} \\ &= W_{cu}(\alpha + 1). \end{aligned} \tag{7.30}$$

Then, losses W_{Lx} can be written as

$$\begin{aligned} &= W_c + W_{cu}\left(\frac{P_x}{P_r}\right)^2 \\ &= \alpha W_{cu} + W_{cu}\left(\frac{P_x}{P_r}\right)^2 \\ &= W_{cu}\left[\alpha + \left(\frac{P_x}{P_r}\right)^2\right] \end{aligned} \tag{7.31}$$

Dividing Eqn. (7.31) by Eqn. (7.30) and using Eqn. (7.29), we have,

$$\frac{W_{Lx}}{W_{Lr}} = \frac{\alpha + (P_x/P_r)^2}{(\alpha + 1)} = \frac{1}{(1 - e^{-N/\tau})} \tag{7.32}$$

Therefore,

$$\frac{P_x}{P_r} = \sqrt{\frac{1 + \alpha}{1 - e^{-N/\tau}} - \alpha} \tag{7.33}$$

The above equation enables us to determine the short time duty rating P_x of the motor for given values of the ratio of constant losses to copper losses at full load (α), heating time constant (τ) and duration of the load (N).

Example 7-3. Determine the half hour rating of a 25 kW motor having a time constant of 1.5 hours. Assume that the motor cools down completely between each load period and that the iron losses which remain constant are 90 per cent of copper losses at full load.

Solution:
$$\frac{P_x}{P_r} = \sqrt{\frac{1+\alpha}{1-e^{-N/\tau}} - \alpha}$$

$$\alpha = 0.9,\ N = 0.5 \text{ hr},\ \tau = 1.5 \text{ hrs}$$

$$P_r = 25 \text{ kW}$$

Therefore,
$$P_x = 25\sqrt{\frac{1+0.9}{(1-e^{-0.5/1.5})} - 0.9}$$

$$= 25\sqrt{\frac{1.9}{(1-0.72)} - 0.9}$$

$$= 25\sqrt{\frac{1.648}{0.28}}$$

$$= 25 \cdot 2.426$$

$$= 60.65 \text{ kW}$$

say, 60 kW.

(ii) *Intermittent rating*: When a motor is intermittently loaded it will cool off during the time it is off and its temperature will rise when it is on as shown in Fig. 7.18. Let $\theta_{h_1}, \theta_{h_2}, \ldots$ be the rise in temperature after heating and $\theta_{c_1}, \theta_{c_2}, \ldots$ be those after cooling. N and τ denote the duration of heating and heating time constant, while R and τ' the duration of cooling and cooling time constant. The temperature θ_m' is the permissible final temperature rise of the motor, which should not exceed that reached when loaded to its continuous full load rating P_r. Let us, first, determine, this maximum temperature reached with intermittent loads.

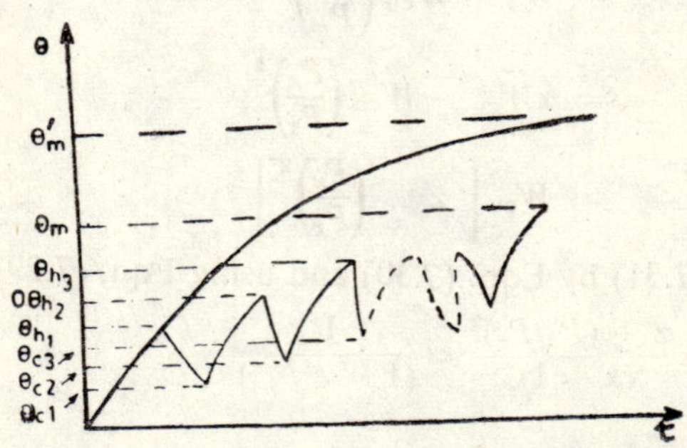

Fig. 7.18. Heating and cooling curves for intermittent Loads.

$$\theta_{h_1} = \theta_m' (1 - e^{-N/\tau}) \tag{7.34}$$

and
$$\theta_{c_1} = \theta_{h_1} e^{-R/\tau'} \tag{7.35}$$

Denoting $-\dfrac{N}{\tau} = x$ and $-\dfrac{R}{\tau'} = y$, we have

$$\theta_{h_1} = \theta_m' (1 - e^x) \text{ and}$$

$$\theta_{c_1} = \theta_{h_1} e^y$$

$$= \theta_m' (1 - e^x) e^y$$

$$\theta_{h_2} = \theta_m' - (\theta_m' - \theta_{c_1}) e^x$$

$$= (1 - e^x) \theta_m' + \theta_{c_1} e^x$$

$$= (1 - e^x) \theta_m' + \theta_m' (1 - e^x) e^y e^x$$

$$= \theta_m' (1 - e^x) (1 + e^x e^y)$$

$$\theta_{c_2} = \theta_{h_2} e^y$$

$$= \theta_m' e^y (1 - e^x) (1 + e^x e^y)$$

$$\theta_{h_3} = \theta_m' (1 - e^x) + \theta_{c_2} e^x$$

$$= \theta_m' (1 - e^x) + \theta_m' e^x e^y (1 - e^x) (1 + e^x e^y)$$

$$= \theta_m' (1 - e^x) \{1 + e^x e^y + e^{2x} e^{2y}\}.$$

For n times intermittency, we have

$$\theta_{hn} = \theta_m' (1 - e^x) \{1 + e^x e^y + e^{2x} e^{2y} + \ldots e^{(n-1)x}\, e^{(n-1)y}\}$$

$$= \theta_m' (1 - e^x) \left\{\frac{1 - e^{nx}\, e^{ny}}{1 - e^x e^y}\right\} \tag{7.36}$$

As $n \to \infty$ both e^{nx} and e^{ny} will be zero as x and y are negative. Since θ_{hn} should not exceed θ_m the maximum permissible temperature rise at continuous rating of P_r, we have

$$\theta_m = \theta_m' (1 - e^x) \frac{1}{1 - e^x e^y}$$

$$= \theta_m' \frac{(1 - e^{-N/\tau})}{\{1 - e^{-(N/\tau + R/\tau')}\}} \tag{7.37}$$

As explained in the earlier case

$$\frac{\theta_m'}{\theta_m} = \frac{W_{Lx}}{W_{Lr}} = \frac{1 - e^{-(N/\tau + R/\tau')}}{(1 - e^{-N/\tau})}$$

$$= \frac{\left\{\alpha + \left(\frac{P_x}{P_r}\right)^2\right\}}{(\alpha + 1)} \tag{7.38}$$

Hence, $$\left(\frac{P_x}{P_r}\right) = \sqrt{\frac{(\alpha + 1)\{1 - e^{-(N/\tau + R/\tau')}\}}{(1 - e^{-N/\tau})} - \alpha} \tag{7.39}$$

Example 7-4. The heating and cooling time constants of a 100 kW motor are 90 and 120 minutes respectively. Find the rating of motor when subjected to a duty cycle of 18 minutes on certain load and 30 minutes on no load. Assume that the losses are proportional to square of load.

Solution: $$\frac{P_x}{P_r} = \sqrt{\frac{(\alpha + 1)\{1 - e^{-(N/\tau + R/\tau')}\}}{(1 - e^{-N/\tau})} - \alpha}$$

Since the losses are proportional to the square of the load, iron losses are negligible, i.e. $\alpha = 0$.

Therefore, $$\frac{P_x}{P_r} = \sqrt{\frac{1 - e^{-(N/\tau + R/\tau')}}{1 - e^{-N/\tau}}}$$

$$P_r = 100 \text{ kW}; \quad N = 18 \text{ min}; \quad \tau = 90 \text{ min};$$
$$R = 30 \text{ min}; \quad \tau' = 120 \text{ min}.$$

Hence, $$P_x = P_r \sqrt{\frac{1 - e^{-(18/90 + 30/120)}}{1 - e^{-18/90}}}$$

$$= P_r \sqrt{\frac{1 - e^{-9/20}}{1 - e^{-1/5}}}$$

$$= P_r \sqrt{\frac{1 - 0.638}{1 - 0.819}} = P_r\sqrt{2}$$

$$= 100\sqrt{2} = 141.4 \text{ kW say, } 140 \text{ kW}.$$

7.4 Effect of Load Inertia

The principal load on motors driving loads of heavy inertia and having frequent starts is the acceleration and braking of the rotating masses. The load torque acting on the motor shaft is practically negligible. Rating of drives of such equipment essentially depends on the variation of torque during acceleration and braking. The power rating of the motor will be proportional to the product of rated torque and maximum speed of the motor. In actual practice, such loads have permissible limits of speed of rotation, torque and acceleration. For example, constructional requirements of excavator specify the angular speed of turning ω_0, angular acceleration A and the moment of inertia of the turning part of the excavator. Hence, it is convenient to determine the rating of the drive starting from the above-mentioned parameters.

While operating with a triangular speed-time curve as shown in Fig. 7.19, the drive races upto the rated speed and immediately decelerates to complete stoppage. Assuming that the torque during both acceleration and braking remains the same, the power rating P_1 is expressed as

$$P_1 = T \cdot \omega_0, \qquad (7.40)$$

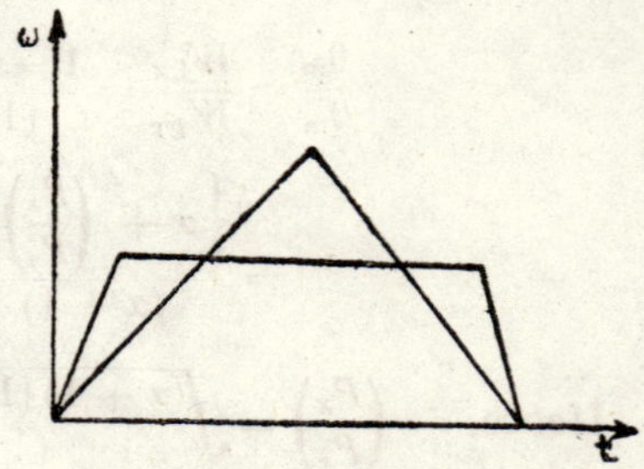

Fig. 7.19. Two typical speed-time curves.

where

$$T = J \cdot \frac{d\omega}{dt} \text{ (neglecting load torque)}$$

$$= J \cdot A \qquad (7.41)$$

With a triangular speed-time curve A can be taken as a constant.

Hence, $$P_1 = J\omega_0 A \qquad (7.42)$$

where P_1 = rating of motor, watts

J = moment of inertia, kg-m²

ω_0 = rated speed, rad/sec.

A = angular acceleration, rad/sec²

Very often, during the operation of excavators or similar equipment used in construction industry, the speed-time curve may be of different shape. If the operation were to follow a trapezoidal speed time curve as shown in Fig. 7.19, the period of steady state motion could be considered as a period of operation with constant load.

The equivalent rating of the motor, then, will be given by

$$P_{eq} = P_1 \sqrt{\frac{N}{N+R}}$$

$$= J\omega_0 A \sqrt{\frac{N}{N+R}}, \tag{7.43}$$

where N and R represent the time for which the motor is on and off, respectively during a cycle.

7.5 Load Equalisation

Certain loads fluctuate very widely within short intervals of time. Electric hammers, presses, reciprocating pumps and steel billet rolling mills belong to such category of loads. When the motor is subjected to heavy loads, it will draw large current from the supply. These sudden peak demands of current, in addition to causing heavy losses, give rise to appreciable voltage drop in the lines and mechanical forces between the conductors carrying the current. If the fluctuating load were to be a predominant one among those connected to the supply busbars, the other loads connected to the same supply lines would be subjected to undesirable voltage fluctuations. Therefore, it is necessary to smoothen out the load fluctuations. This process is called load equalisation.

Load equalisation is often achieved by means of a flywheel, which is mounted on the motor shaft, if the speed of the motor is not to be reversed. In case of reversing rolling mills, the flywheel is mounted on the motor-generator set feeding the driving motor. In order that the flywheel operates effectively, the driving motor should have a drooping speed characteristic. During intervals of heavy load, speed of the motor decreases. This enables the flywheel to release a portion of the stored kinetic energy, which together with the energy drawn from supply, will meet the demand of load. During off peak load period, the motor draws more energy than required by the load. The surplus energy taken is again stored as kinetic energy in the flywheel, whose speed, then, increases. The load torque required and motor torque developed as well as the speed variations with time are depicted in Fig. 7.20.

7.5.1 Determination of the Moment of Inertia of the Flywheel

The general equation of motion is written as

$$T - T_L = J \cdot \frac{d\omega}{dt} \tag{7.44}$$

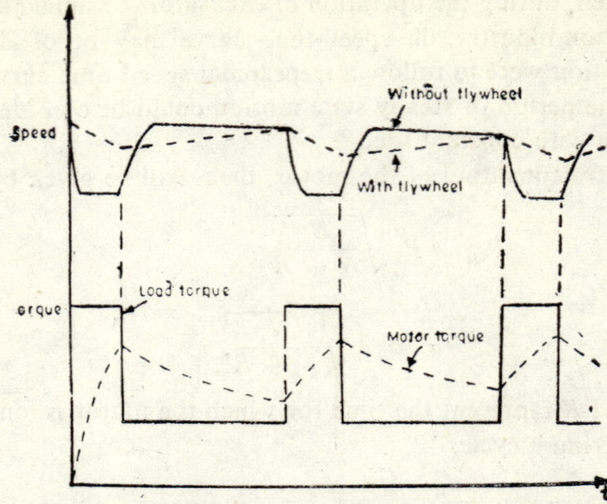

Fig. 7.20. Variations of speed, load torque and motor torque against time.

For successful operation of the flywheel mechanism, the motor speed should vary with torque and let us assume that the variation is linear. Hence,

$$\omega = \omega_0 - \frac{\omega_0 - \omega_r}{T_r} \cdot T, \tag{7.45}$$

where ω, ω_0 and ω_r denote instantaneous speed, no-load speed and rated speed. T and T_r represent the instantaneous and rated torques. Therefore,

$$\frac{d\omega}{dt} = -\frac{\omega_0 - \omega_r}{T_r} \cdot \frac{dT}{dt} \tag{7.46}$$

$$J \cdot \frac{\omega_0 - \omega_r}{T_r} \cdot \frac{dT}{dt} + T = T_L \tag{7.47}$$

i.e.,

$$t_m \frac{dT}{dt} + T = T_L, \tag{7.48}$$

where

$$t_m = J \cdot \frac{(\omega_0 - \omega_r)}{T_r} \tag{7.49}$$

denotes the mechanical time constant in seconds.

Equation (7.47) clearly indicates that at any instant a portion of the torque demand of the load is met by the developed torque of the motor, while the rest is contributed by the flywheel.

The instantaneous torque developed by the motor during either heavy load period or light load period can be obtained by solving Eqn. (7.48). It will be of the form

$$T = T_L (1 - e^{-t/t_m}) + T' e^{-t/tm}, \tag{7.50}$$

where T' is the torque developed by the motor at the instant of application or removal of the load.

Using Eqn. (7.50), the torque developed by the motor at the end of the heavy load (T_{L_h}) acting for a time interval, t_h, is written as

$$T_{max} = T_{L_h}(1-e^{-t_h/t_m}) + T_{min}\, e^{-t_h/t_m}, \tag{7.51}$$

where T_{min} is the motor torque at the instant when heavy load T_{L_h} is applied.

Similarly, the motor torque at the end of the light load (T_{L_1}) acting for a time interval, t_1, is given by,

$$T_{min} = T_{L_1}(1-e^{-t_1/t_m}) + T_{max}\, e^{-t_1/t_m} \tag{7.52}$$

From Eqns. (7.51) and (7.52), we have

$$e^{-t_h/t_m} = \frac{T_{L_h}-T_{max}}{T_{L_h}-T_{min}} \tag{7.53}$$

$$e^{-t_1/t_m} = \frac{T_{min}-T_{L_1}}{T_{max}-T_{L_1}} \tag{7.54}$$

Also, from Eqn. (7.49)

$$J = T_r \cdot \frac{t_m}{(\omega_0-\omega_r)} \tag{7.55}$$

Using Eqns. (7.53) and (7.54),

$$t_m = \frac{t_h}{\log_e\left(\dfrac{T_{L_h}-T_{min}}{T_{L_h}-T_{max}}\right)} \tag{7.56}$$

$$\text{or} \quad \frac{t_1}{\log_e\left(\dfrac{T_{max}-T_{L_1}}{T_{min}-T_{L_1}}\right)} \tag{7.57}$$

Hence, moment of inertia of flywheel

$$J = \frac{T_r}{(\omega_0-\omega_r)} \cdot \frac{t_h}{\log_e\left(\dfrac{T_{L_h}-T_{min}}{T_{L_h}-T_{max}}\right)} \tag{7.58}$$

$$\text{or} \quad \frac{T_r}{(\omega_0-\omega_r)} \cdot \frac{t_1}{\log_e\left(\dfrac{T_{max}-T_{L_1}}{T_{min}-T_{L_1}}\right)} \tag{7.59}$$

Knowing the radius of gyration, the weight of the flywheel to be used for load equalisation can be easily determined.

Example 5-7. A 3-phase, 50 kW, 6-pole, 960 rpm induction motor has a constant load torque of 300 N·m and at wide intervals additional torque of 1500 N-m for 10 seconds. Calculate (i) the weight of the flywheel used for load equalisation, if the motor torque were not to exceed twice the rated torque and the radius of gyration is 0.9 m, (ii) the time taken after removal of additional load before the motor torque becomes 700 N-m.

Assume that the induction motor operates on that portion of the slip torque characteristic, which is linear.

Solution: Rated torque $T_r = \dfrac{50\cdot 1000\cdot 60}{2\pi\cdot 960} = 497.36$ N-m

$$T_{max} = 2\cdot 497.36 = 994.72 \text{ N-m}$$

$$T_{min} = 300 \text{ N-m}, \ T_{L_h} = 1500+300 = 1800 \text{ N-m}$$

$$t_h = 10 \text{ sec.}$$

$$\omega_0 = \frac{2\pi.1000}{60}, \ \omega_r = \frac{2\pi\cdot 960}{60}$$

$$J \quad \frac{T_r}{(\omega_0-\omega_r)}\cdot\frac{t_h}{\log_e\left(\dfrac{T_{L_h}-T_{min}}{T_{L_h}-T_{max}}\right)}$$

$$= \frac{497.36}{\dfrac{2\pi}{60}(1000-960)}\cdot\frac{10}{\log_e\left(\dfrac{1800-300}{1800-994.72}\right)}$$

$$= 2009.65 \text{ kg-m}^2$$

Wt. of flywheel $= J/r^2 = 2009.65/0.81$

$= 2481$ kg.

(ii) Let the time taken be t_1 sec.

$$T_{max} = 994.72, \qquad T_{min} = 700$$

$$T_{L_1} = 300$$

$$J = \frac{T_r}{(\omega_0-\omega_r)}\cdot\frac{t_1}{\log_e\left(\dfrac{T_{max}-T_{L_1}}{T_{min}-T_L}\right)}$$

$$2009.65 = \frac{497.36}{2\pi/60\,(1000-960)}\cdot\frac{t_1}{\log_e\left(\dfrac{994.72-300}{700-300}\right)}$$

$$t_1 = 8.87 \text{ sec.}$$

7.6 Environmental Factors

A 100-kW motor of intermittent rating might be rated for 200-kW, if continuously operated at the North Pole at an ambient temperature of —80°C, since all the heat generated would still not be sufficient to overheat the motor under such ambient conditions.

At high altitudes the reduced air density decreases the cooling effect. The limitation is negligible for elevation less than 1000 m above sea level. For heights greater than 1000 m, the rating of the machine should be reduced according to the figures in the table below:

Height above sea level, m	0–1000	2000	3000	4000
Output for motor speed upto 1000 rpm, per cent	100	95	90	85
Output for motor speeds above 1000 rpm, per cent	100	92	85	77

Totally enclosed machines without auxiliary forced ventilation, which do not permit ventilation and replacement of internal air, do not have as high a rating as similar machines, which are not totally enclosed and which are ventilated in such a manner that fresh air is drawn across the stator and rotor windings.

PROBLEMS

1. Assuming that the temperature increases according to an exponential law, determine the heating time constant for the motor, when its temperature rise after 30 minutes of operation is 60 per cent of the final value.
2. The temperature rise of a motor after operating for 30 minutes on full load is 20°C, after another 30 minutes on the same load the temperature rise becomes 30°C. Assuming that the temperature increases according to an exponential law, determine the final temperature rise and the time constant.
3. A motor on a heat run test gave the following readings of mean temperature:

Time (hr)	0	0.25	0.5	0.75	1.0	1.25	1.5	1.75
Temperature (°C)	42.3	45.0	47.4	49.5	51.4	53.0	54.4	55.6

Find graphically (a) the final steady temperature and temperature rise and (b) heating time constant. Assume that the room temperature is 30°C.

4. Based on the rms torque, estimate the kW rating of a 750 rpm motor used for driving an equipment having the following load torque curve:
   (i) For the first 10 seconds, the torque is constant at 40 kg-m.
   (ii) For the next 30 seconds, the torque varies linearly with time from 35 kg-m to 15 kg-m.
   (iii) For the last 50 seconds, the torque is constant and equal to 10 kg-m.
5. A motor driving a mining equipment has to supply a load rising uniformly from zero to a maximum of 1500 kW in 20 seconds during acceleration period, 1000 kW for 50 seconds during the full load period and during deceleration period of 10 seconds when regenerative braking takes place, the kW returned to the mains falls from an initial value of 500 to zero uniformly. The interval for decking before the next load cycle starts is 20 seconds. Estimate a suitable kW rating of the motor, based on rms power.
6. The duty cycle of a motor driving a grab bucket hoist for unloading coal from a barge into a bunker is as follows:

Operation	Closing of bucket	Hoisting	Opening of bucket	Lowering of bucket	Rest
Duration (sec)	5	10	3	10	15
Power required (kW)	50	80	40	60	0

(a) Using the rms method, determine the continuous rating of the motor.
(b) Quotations are available from two motor manufacturers to supply the required motor. They include the following efficiency-load data and prices:

	Efficiency vs. load per unit					*Price in Rs.*
Motor	0.25	0 50	0.75	1.0	1.25	
A	0.824	0.895	0.893	0.87	0.858	25,000
B	0.803	0·876	0.893	0.896	0.893	30,000

The average net cost of energy is 40 paise/KWh. The total fixed charges on invested capital are 20 per cent. The drive will be working on an average for 2200 hr/annum. Which of the above two motors would you choose?

7. The motor of problem 2 above has, on its rated continuous load of 10 kW, copper losses equal to iron losses. Estimate its 1 hour rating.
8. The 15 minutes rating of a motor used in domestic mixer is 400 watts. If the heating time constant is 60 minutes, determine the continuous rating, if maximum efficiency of motor occurs at 80 per cent of full load.
9. On the basis of heating, select a suitable motor for the following intermittent duty:

   (i) Constant load of 35 kW for 3 sec,
   (ii) Constant load of 15 kW for 20 sec,
   (iii) Constant load of 35 kW for 2 sec. and
   (iv) Constant load of 10 kW for 15 sec.

   Between the operating periods (ii) and (iii) there is a period of rest for 37 seconds and after (iv) there is another rest period for 43 sec.
10. A 8-pole, 50 Hz, three phase induction motor has a flywheel of 1500 kg-m^2 moment of inertia. Load torque is 200 kg-m for 10 sec. No load period is long enough for the flywheel to regain its full speed. Motor has a slip of 4 per cent at a torque of 100 kg-m. Calculate

    (i) the maximum torque developed by the motor, and
    (ii) the speed at the end of the deceleration period.

    Assume that the speed torque curve of the motor is linear in the operating range.

Answers

1. 32 mins
2. 43.28 mins; 40°C
3. (a) 64.5°C. 34.5°C (b) 1.93 hrs.
4. 16.71 kW
5. 811.4 kW
6. (a) 52.22 kW
7. 12.9 kW
8. 154 W
9. 10.05 kW
10. (i) 1715 Nm; (ii) 697.5 rpm

Chapter 8

Introduction to Solid State Controlled Drives

The advent of thyristors capable of handling large currents has revolutionized the field of electric power control. Thyratrons, ignitrons, mercury are rectifiers, magnetic amplifiers, motor-alternator sets have all been replaced by solid state circuits using diodes and thyristors. Thyristor controlled drives using both dc and ac motors find wide application in industry as variable speed drives. Changing from electromechanical and other form of control to electronic control, generally, has the advantages of higher accuracy, better reliability and response and higher efficiency.

8.1 DC Motor Systems

Solid state control circuits have been developed corresponding to each one of the methods of modifying the speed torque characteristics of a dc motor, viz., by adjusting the armature voltage or by adjusting the field current or by both.

Adjustable armature voltage can either be obtained from controlled rectifier circuits (usually termed as converters) or from chopper circuits. The latter are, normally used when dc supply is readily available. While motor voltage control is achieved in converter circuits by varying the phase angle at which the thyristors are fired relative to the applied alternating voltage waveform, it is obtained in chopper circuits by changing the on-to-off time ratio for which the dc supply voltage is applied to the motor. Instead of converter circuits, it is possible to use an uncontrolled rectifier, which gives a constant direct voltage, followed by a chopper to provide a variable mean direct voltage output.

Variation in field current can be obtained by supplying the field winding from an auxiliary controlled rectifier circuit.

8.1.1 Controlled Rectifier Circuits

Rectifier circuits are classified according to the number of output voltage pulses in a period of the ac line power frequency. The simplest converter is of $m=1$, where m equals the number of voltage pulses per cycle of the supply voltage. The basic forms of these converters are shown in Figs. 8.1-8.3. Fig. 8.1(a) shows a dc motor fed by such a converter. During standstill conditions of the motor, when S_1 is more positive than S_2, the

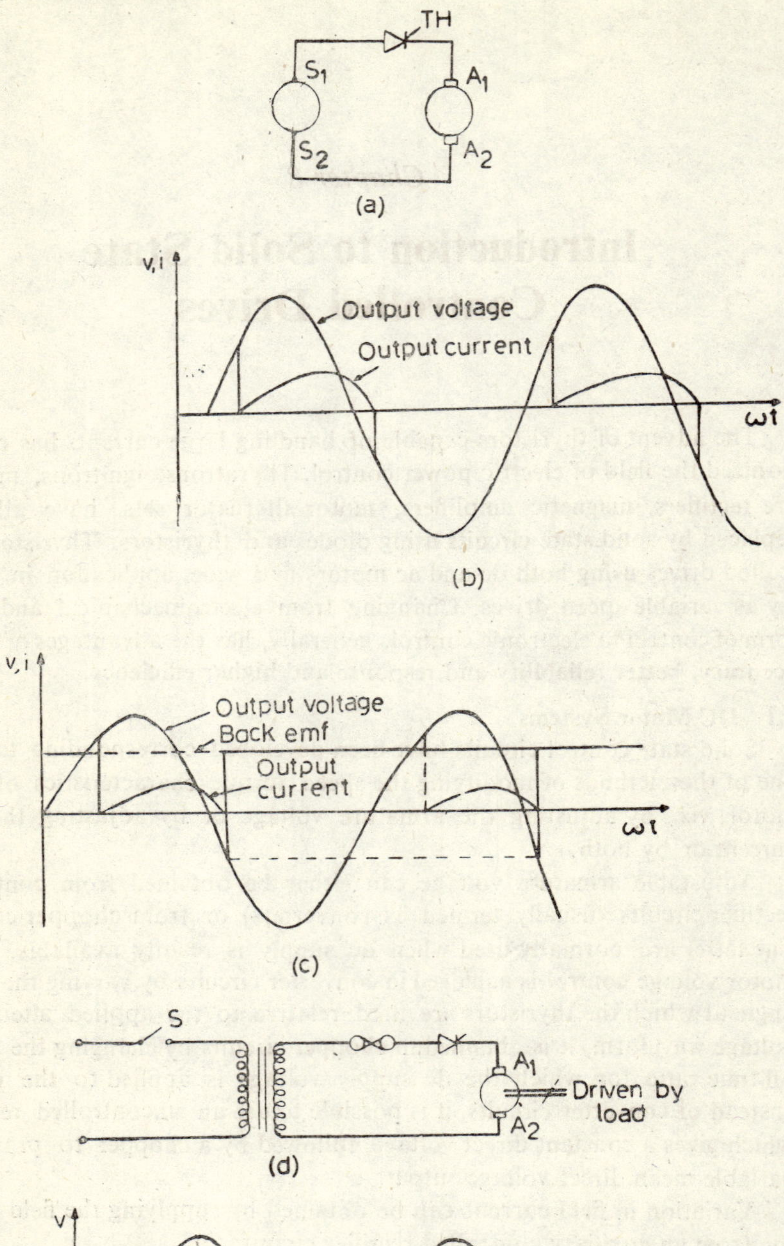

Fig. 8.1. **Operation of *1-pulse* converter.**

thyristor TH may be gated on, producing the output voltage and current waveforms shown in Fig. 8.1(b). That is, during the positive half cycle of the supply voltage waveform, the current flows from the source into the motor armature circuit. At the start of the negative half cycle, due to the reverse bias to the thyristor, the current would stop if the armature circuit were purely resistive. Since the armature circuits have appreciable inductance, during the negative half cycle although S_2 becomes more positive than S_1, current continues to flow from the source into the armature until the stored inductive energy is dissipated.

During running conditions of the motor, the thyristor can start conducting only when the instantaneous value of the source voltage becomes equal to or exceeds the induced emf in the armature and when it receives a gating pulse. The thyristor reaches its blocking state when the armature inductance, has discharged its stored energy and the current falls to zero, as during standstill conditions of the motor. The presence of the induced emf, however, causes a different output voltage waveform [Fig. 8.1(c)].

If the motor were driven by a load connected to its shaft such that terminal A_2 is positive with respect to terminal A_1, power would flow from the motor armature to the ac source. This mode of operation is known as inverting. If during operation the ac supply should fail (indicated by opening the switch S in Fig. 8.1(d), there would be no ac source to commutate the dc current. That is, due to the small circuit impedance; when S opens, the armature current continues to flow and increase until the fuse or the thyristor opens or the motor fails. Another possibility of failure during inverting operation exists when the magnitude of the induced emf of the motor becomes too large with respect to the ac line voltage as shown in Fig. 8.1(e). The thyristor begins to conduct at point A. At point B the ac line voltage and the induced emf impress a negative anode voltage on the thyristor, but due to the inductance of the motor armature the current continues to flow. At point C the ac line voltage and the induced emf produce a positive anode voltage and the thyristor continues to conduct. The armature current continues to increase during successive half-cycle periods, until a circuit failure occurs.

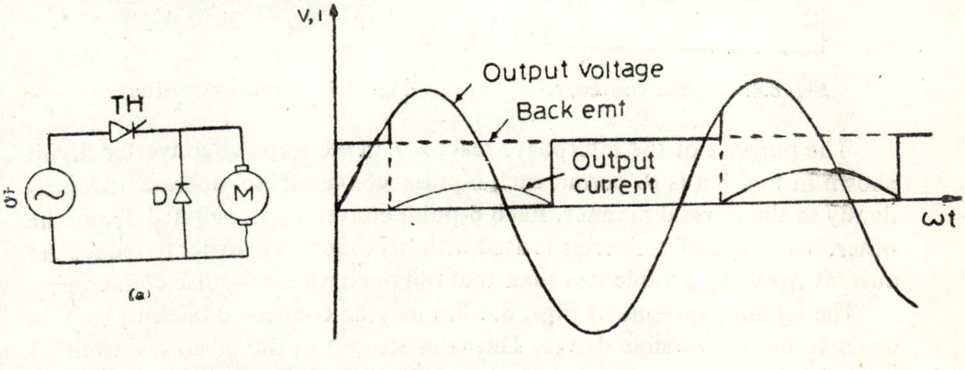

Fig. 8.2. 1-pulse converter with freewheeling diode.

Figure 8.2(a) shows another form of a single pulse converter, having a free-wheeling diode. This circuit, because of the presence of the freewheel ing diode, cannot feed energy back into the ac line. The corresponding output voltage and current waveforms are shown in Fig. 8.2 (b). During the negative half cycle, thc energy stored in the armature inductance flows from the armature through the freewheeling diode at negligibly small voltage. Hence, the conduction of current during the negative half-cycle is for a much longer period than in Fig. 8.1 (c). The form factor (ratio of rms to average value) of the armature current is lower than in Fig. 8.1 (c).

Figure 8.3 is a typical $m = 1$ reversing system. Single pulse converters are used only with motors of rating less than 3 kW, since they have a higher form factor for the armature current than the other types of rectifiers. Yet another disadvantage is that a transformer is required to eliminate the direct components of voltage and current from the ac line. The transformer for use with single pulse converter fed motors must be specially designed to handle the dc component.

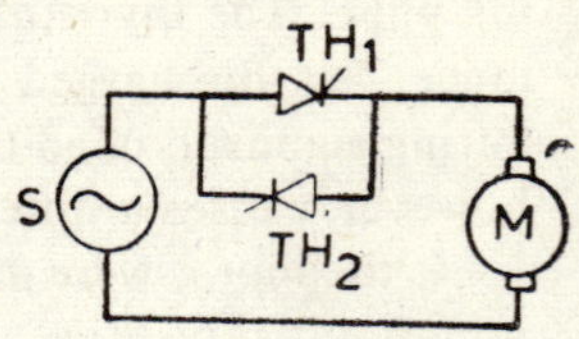

Fig. 8.3. Reversible 1-pulse converter.

Three phase supply voltages are used for dc drive systems of about 5k W and larger. Three phase rectifier circuits give more number of voltage pulses per cycle of supply frequency, thus ensuring the flow of armature current over a longer portion of the cycle, increasing the form factor and thereby reducing the heating of the armature. Also, the power is drawn from a three phase system which generally has more capability of supplying power than a single phase system.

The commonly used three phase rectifiers for dc motor drives are shown in Fig. 8.4 ($m = 3$), Fig. 8.5 ($m = 6$) and Fig. 8.6 ($m = 12$).

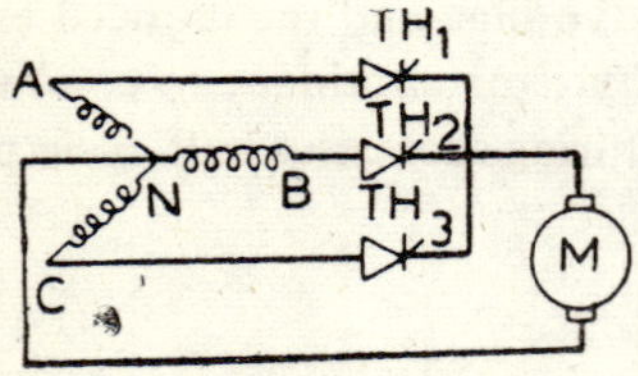

Fig. 8.4. 3-pulse converter.

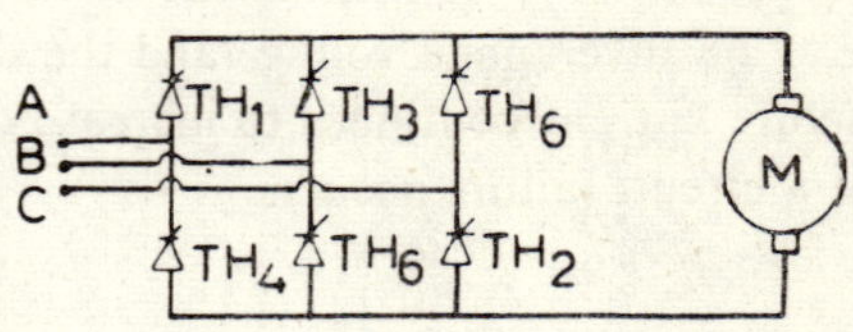

Fig. 8.5. 6-pulse converter.

The purpose of the interphase reactor L in the 12-pulse converter circuit shown in Fig. 8.6 is to permit each 6-pulse converter to operate independently in the normal manner. Each 6-pulse converter is isolated from the other. This type of converter is used with large motors in order to reduce the current ripple to a value less than that obtained with a 6-pulse converter.

The rectifier systems of Figs. 8.4-8.6 may be connected back to back in order to have reversible drives. This connection ties the positive output of one rectifier to the negative output of the other. This allows motor

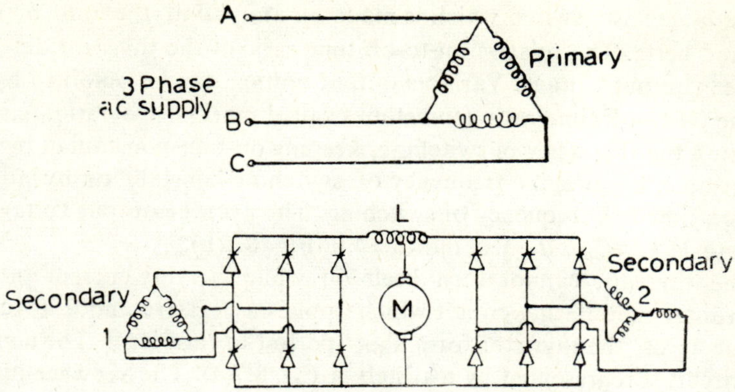

Fig. 8.6. 12-pulse converter.

armature current to flow in either direction. The problem that must be overcome with this type of connection is to prevent both sides of the converter from coming on at the same time and, hence, shorting the ac line. The commonly used techniques to limit the short circuit current are: (i) insertion of a reactor in the armature circuit, (ii) provide time delay before turning on the opposite rectifier and (iii) introduction of a current sensing device which prevents one side from turning on, if current were flowing in the opposite side.

8.1.2 *Chopper Circuits*

Chopper circuits are used to control the speed of dc motors which are fed from fixed voltage sources like batteries or uncontrolled rectifiers. Fig. 8.7(a) shows the basic chopper circuit that supplies variable direct voltage to a series motor. The input to the chopper is a fixed direct voltage *V*. The

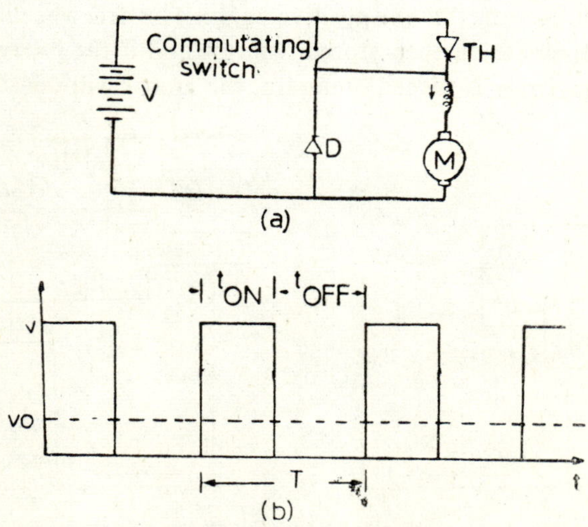

Fig. 8.7. Chopping of voltage.

thyristor acts as a switch which is made on and off at the rate of several hundred hertz. The relative on-to-off time ratio of the thyristor determines the mean output voltage. Variable output voltage can be obtained by varying the on-to-off time ratio (sometimes called mark-space ratio) either by adjusting the frequency of switching, keeping on-time constant or by adjusting on-time, keeping the frequency of switching constant or by adjusting both on-time and frequency of switching. The average output voltage V_0 is equal to $V \cdot t_{on}/(t_{on} + t_{off})$ as indicated in Fig. 8.7(b).

The thyristor cannot turn itself off while carrying current and hence requires a commutating circuit which applies a negative anode to cathode voltage across the thyristor for a short-period to turn it off. The commutating circuit is represented by a switch in Fig. 8.7(a). The freewheeling diode provides a path for the armature current to flow while the thyristor is not conducting.

Three chopper circuits commonly used for obtaining variable direct voltage from fixed voltage source are described below:

The Morgan chopper: This circuit, shown in Fig. 8.8, is typical of class *B* commutation, i.e , self-commutation by a resonating circuit helped by a saturable reactor. The greatest advantage of this circuit is that it needs only one thyristor. Therefore, the on- time is fixed by LC parameters and the average voltage across the motor is varied by adjusting the chopping frequency. The gating oscillator generates variable frequency.

When the thyristor TH is fired, the capacitor C, having a positive polarity at the dot in the figure discharges around the circuit formed by C-TH and L to acquire negative polarity. As the current again reverses, the voltage across the reactor is held for some time and then saturation occurs such that the entire capacitor voltage appears across the thyristor. The thyristor gets reverse biased and if the discharge current is more than the thyristor current, it turns off. The capacitor continues to carry load current until it charges up fully with the dot positive again. The freewheeling diode offers a path to dissipate further stored energy and if the energy is dissipated before the thyristor is turned on again, the motor will coast.

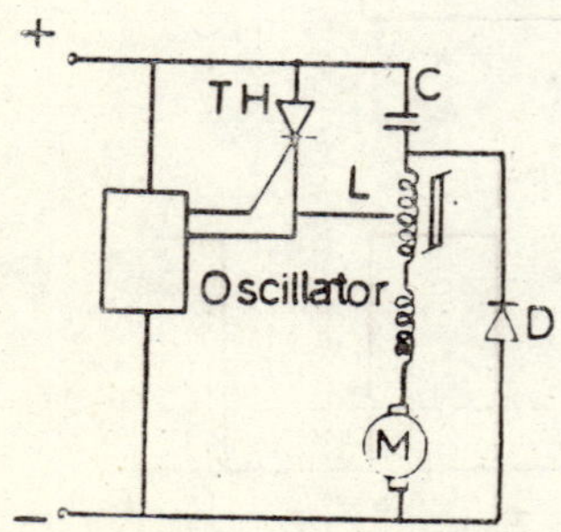

Fig. 8.8. Morgan chopper.

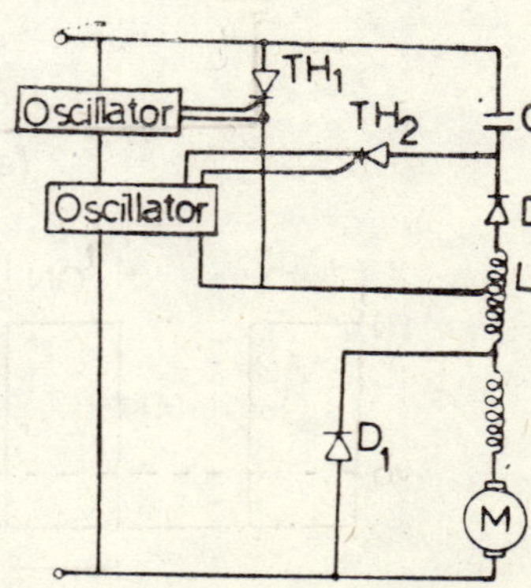

Fig. 8.9. Jones chopper.

The Jones chopper: Figure 8.9 shows the circuit, which is characterized by class D commutation. A charged capacitor C, switched by an auxiliary thyristor TH_2 and the autotransformer T constitute the commutating circuit. Although, in principle, both the on-time and off-time can be varied due to the presence of TH_2, usually, as indicated in the figure, the off-time or period T is varied by means of the gating oscillator of TH_1, keeping on-time fixed by means of the oscillator of TH_2.

When thyristor TH_1 is fired, the capacitor charged positively at the dot discharges around the circuit formed by C — TH_1 — L and D and reverses polarity. The diode D prevents further oscillation of the resonating LC circuit. Hence, the capacitor retains its charge until TH_2 is switched on. Then, the discharge causes TH_1 to be reverse biased and turns it off. The capacitor charges up with the dot positive again and TH_2 turns off, since the current through it falls below the holding value when the capacitor is recharged. The process repeats itself when TH_1 is switched on again.

Even if the capacitor had not charged fully by the time TH_1 was switched on again, no damage would be done, since the load current ensures that the induced emf in L gives the capacitor sufficient commutating energy. Due to this, the thyristors have to be rated at higher voltages.

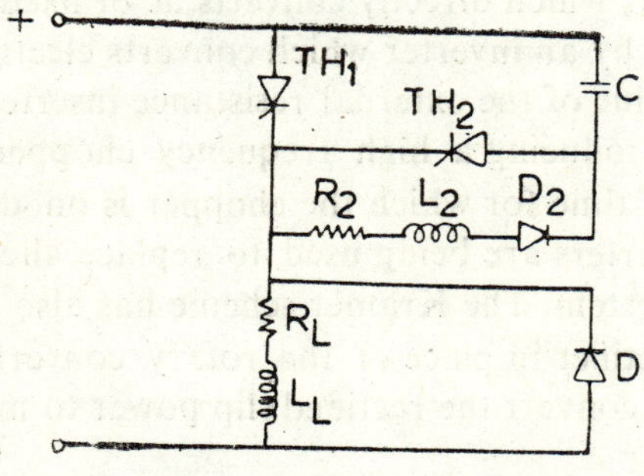

Fig. 8.10. Oscillation chopper.

The oscillation chopper: This circuit derives its name from the resonating or oscillating nature of the class D commutation Fig. 8.10 shows the basic circuit, which differs from Morgan and Jones Chopper in that there is neither a saturable reactor nor an autotransformer in the load circuit.

TH_2 must be triggered first so that C may get fully charged with the dot positive. Then, when TH_1 is switched on, so that load current can flow, C reverses polarity through the resonating circuit C—TH_1—R_2—L_2 and D_2. The capacitor remains in that charged condition because of D_2 until TH_2 is fired. This discharges C, causes TH_1 to be reverse biased and turns it off.

In choosing one power control circuit over the other, the following points should be borne in mind.

(1) Power supply: (a) dc (b) ac—single phase or three phase.

(2) Form factor for the motor: (a) rectifier type (b) need of a choke.

(3) Need for inversion.
(4) Need for reversal.
(5) Necessity of a transformer either because of ac line voltage available or rectifier used.
(6) System cost: (a) transformer (b) choke (c) rectifier (d) motor.
(7) Electrical energy costs: (a) rectifier power factor (b) rectifier efficiency.
(8) Amount of ac line pollution caused by the rectifier.

8.2 AC Motor Systems

As described in Chapter 4, the speed-torque characteristics of an induction motor can be controlled by adjusting the stator voltage, by adjusting the frequency of the source, by adjusting the resistance of the rotor circuit and by injecting slip frequency emfs to the rotor circuit. The speed of a synchronous motor can also be controlled by varying the frequency of the voltage applied to the stator.

Solid state control units corresponding to each one of the above control techniques have been developed. Stator voltage variation is accomplished by means of ac regulators, which vary the rms value of the ac voltage applied to a motor by introducing thyristors connected back-to-back in each supply line. Variable frequency power is obtained either by means of a cycloconverter, which directly converts ac of fixed frequency to variable frequency ac or by an inverter which converts electric power from dc to ac. The effective value of the external resistance inserted in the rotor circuit can be varied by introducing a high frequency chopper across the resistance and varying the time for which the chopper is on during the cycle. Static frequency converters are being used to replace the auxiliary machines in the Scherbius system. The Kramer scheme has also been altered by using a diode bridge rectifier in place of the rotary converter, but a dc motor is still required to convert the rectified slip power to mechanical power.

8.2.1 AC Regulators

AC regulators are used to contiol the rms value of the ac voltage applied to motor. There are two methods of control : on off control and phase control.

In on-off control, the thyristors are employed as switches to connect the motor to the source for a few cycles of the source voltage and then to disconnect it for a comparable period. The thyristor, thus, acts as a high speed contactor. This method is known as integral cycle control.

In phase control, the thyristors are used as switches to connect the motor to the supply for a chosen portion of each cycle of the supply voltage. The power circuit configurations for on-off control and phase control do not differ in any manner.

Smooth variation of three-phase ac voltage can be realized by means of different configurations of the power circuit. Circuits of three phase half wave and full wave ac regulators for delta connected motors or star connected motors in which the neutral point is not inaccessible are shown in Fig. 8.11(a) and (b). Either regulator shown in the figure could, of course, be used with either connection of the motor. The half wave controller, shown in Fig. 8.11 (a) effects a saving in the cost of semiconductor devices and does not give rise to dc components in any part of the system. However, it introduces more harmonics into the line current than does the full wave regulator. As regards the heating of the motor windings are concerned, star connected motor fed through a full wave ac regulator is preferred to a delta connected motor fed from a full wave ac regulator. This is so, because any third harmonic voltages developed by the motor back emf can cause circulating currents, when the motor is delta connected.

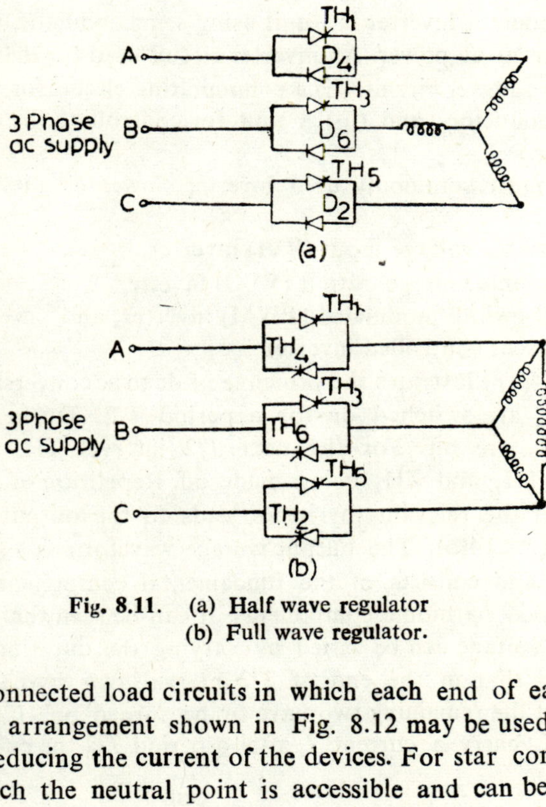

Fig. 8.11. (a) Half wave regulator
(b) Full wave regulator.

For delta connected load circuits in which each end of each phase is accessible, the arrangement shown in Fig. 8.12 may be used and has the advantage of reducing the current of the devices. For star connected load circuits in which the neutral point is accessible and can be opened, the

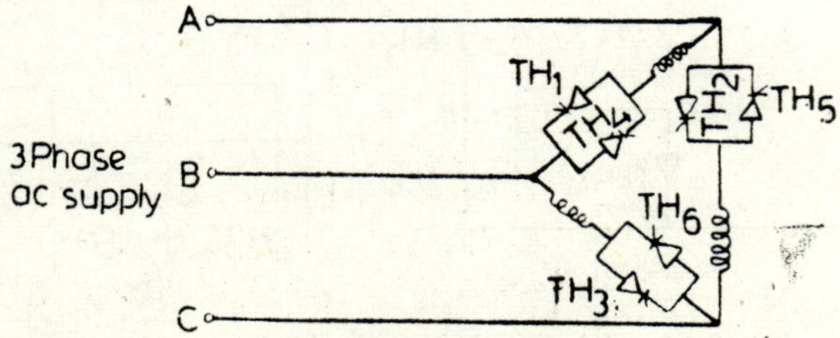

Fig. 8.12. Delta connected regulator.

regulator circuit shown in Fig. 8.13 reduces the number of thyristors required to three and considerably simplifies the control circuitry.

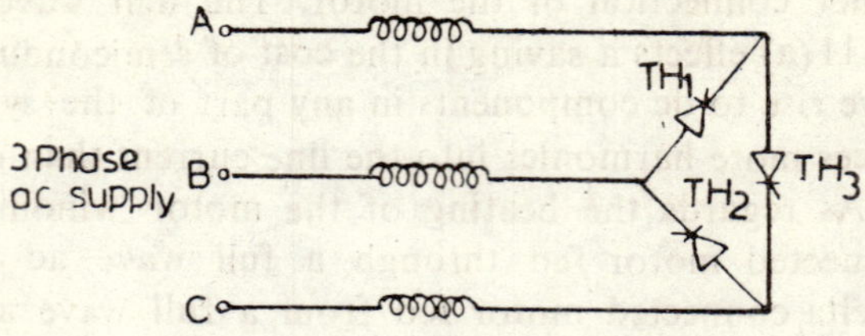

Fig. 8.13. Neutral point regulator.

8.2.2 *Inverters*

A semiconductor inverter is a unit using semiconductor devices to convert dc power to ac power. An inverter circuit used for driving ac motors consists of (i) a power circuit, (ii) a commutating circuit for turning off the power semiconductors and (iii) a unit for controlling the output voltage of the inverter.

The four most commonly used inverter power circuits for ac motor drives are:

(1) the variable voltage input (VVI) inverter,
(2) the variable voltage output (VVO) inverter,
(3) the pulsewidth modulated (PWM) inverter, and
(4) the current controlled inverter.

Figure 8.14(a) illustrates the principle of dc to ac conversion. Thyristors TH_1 and TH_4 are switched on for a period $T/2$. During this interval TH_2 and TH_3 are off. For the next $T/2$ interval TH_2 and TH_3 are switched on, TH_1 and TH_4 being made off. Repetition of such switching on and off of the relevant thyristors leads to the output voltage wave shown in Fig. 8.14(b). The output voltage waveform is a square wave of amplitude E and consists of the fundamental component of frequency $1/T$ and all odd harmonic components. It can be seen that the frequency of the output voltage can be varied by varying the duration $T/2$. It may also be noted that at the end of $T/2$ interval, two thyristors have to be turned off and the remaining two have to be turned on. If, for example, TH_1 and TH_4 carry a current I, after a period $T/2$ the current has to be

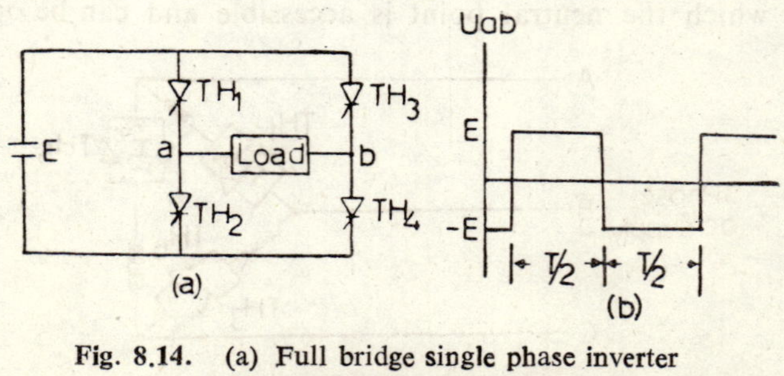

Fig. 8.14. (a) Full bridge single phase inverter
(b) Waveform of load voltage.

brought down to zero by forcing a reverse current greater than *I* through them. This is achieved by an external circuit called the commutation circuit consisting of L and C. Its function is not only to force a reverse current through the conducting thyristor whenever it has to be turned off, but also it should reverse bias the thyristor for a period of time greater than the thyristor turn off time. The above process of turning off the thyristor is called forced commutation and if the duration of this process is very small as compared to $T/2$ interval, it is known as impulse commutation. Such a commutation circuit has hardly any effect on the load voltage waveform.

Figures 8.15 and 8.16 show two simplified versions of the VVI three phase inverter. In Fig. 8.15 a 6-pulse bridge converter is employed to convert the normal three phase ac power to dc power at variable direct voltages through the LC filter. Figure 8.17 shows the voltage waveforms V_{AN}, V_{BN}, V_{CN} and the line-to line voltages V_{AB}, V_{BC} and V_{CA}. Each line voltages waveform is displaced in time phase by 120° electrical degrees from one another and is of quasi-square wave with 120° width.

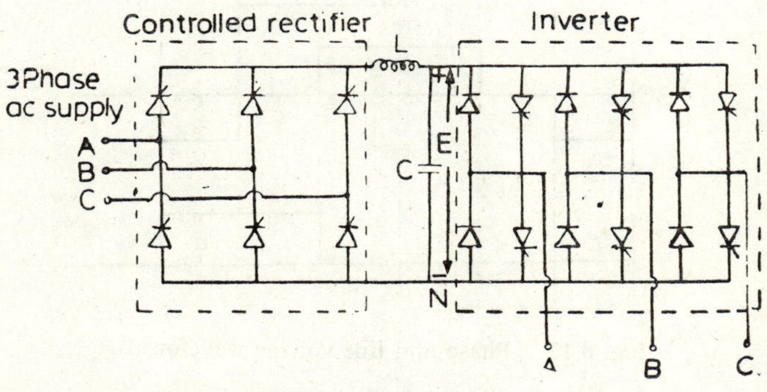

Fig. 8.15. Variable voltage input converter.

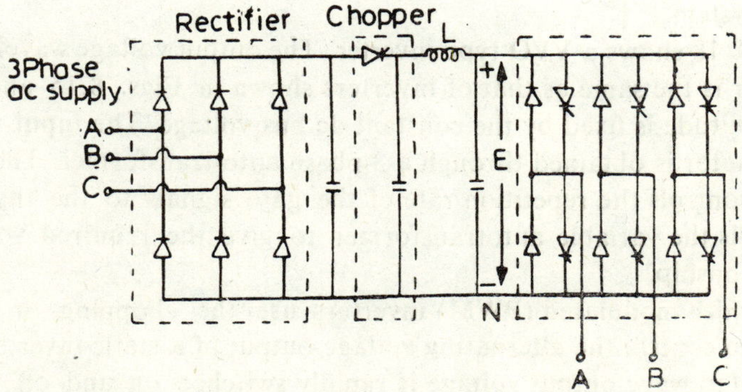

Fig. 8.16. Variable voltage input inverter.

The VVI power circuit shown in Fig. 8.16 is similar to that shown in Fig. 8.15; the 6-pulse controlled bridge rectifier has been replaced by a

6-pulse uncontrolled bridge rectifier and a chopper performs the function of adjusting the output voltage from the diode bridge. This circuit has the ability of feeding from the same dc bus several inverters using one main uncontrolled rectifier.

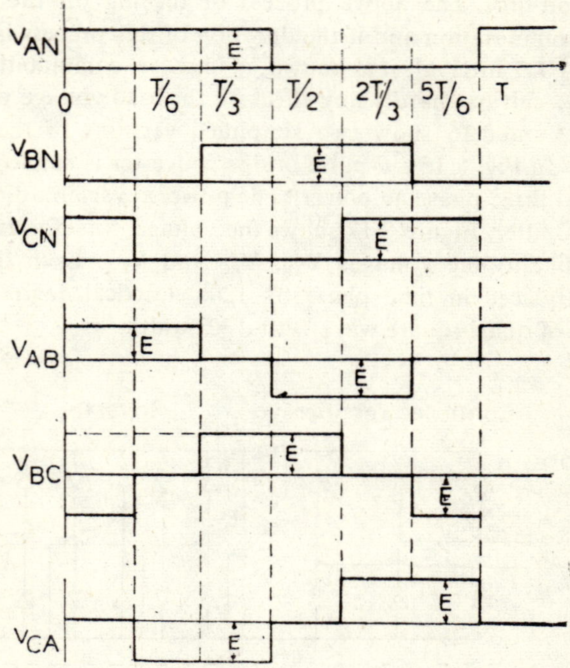

Fig. 8.17. Phase and line voltage waveforms.

The inverters mentioned above cannot return power back to the ac supply lines unless another phase controlled rectifier is added to form a reversing system.

Figure 8.18 shows a VVO type inverter. The output voltage waveform of the inverter is the same as that of inverters shown in Figs. 8.15 and 8.16, but its amplitude is fixed by the constant dc bus voltage. The input voltage to the ac motor is obtained through a 3-phase auto-transformer. The regulator that controls the repetition rate of the gate signals to the thyristors also controls the variable autotransformer to give the required volts per hertz relationship.

Pulse width modulated (PWM) inverters use the chopping or pulsing technique to control the alternating voltage output of a static inverter. The square or step wave output voltage is rapidly switched on and off several times during each half cycle so that a number of pulses of equal amplitude are formed. Each pulse has the amplitude of the inverter input voltage V_{dc}. The magnitude of the fundamental output voltage is controlled by variation of the total on-time during a half cycle. The PWM technique is easily

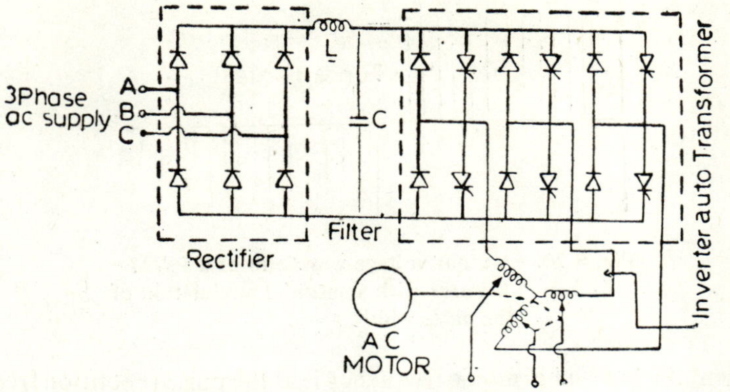

Fig. 8.18. Variable voltage output inverter.

applied to the single phase bridge inverter in Fig. 8.14(a), since each side of the inverter can commutate independently and an interval of zero voltage is obtained whenever the two load terminals are conncted to the same dc line. By commutating one side of the bridge several times during a half-cycle, the output voltage waveform of Fig. 8.19(a) can be obtained. Simple PWM inverters are easily fabricated to generate a waveform with only two pulses per half cycle within a six step envelope, as shown in Fig. 8.19(b). Significant fifth and seventh harmonics are present in this waveform and they cause appreciable deterioration in the low speed performance of the ac motor. In order to eliminate the low order harmonics, more refined techniques of PWM are used, in which high frequency pulsing occurs throughout the half-cycle.

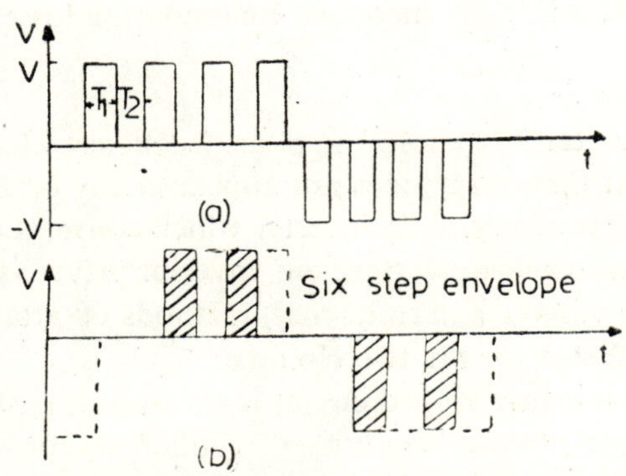

Fig. 8.19. Basic waveforms of PWM.

In sophisticated PWM systems, the pulse width is varied throughout the half-cycle in a sinusoidal manner (Fig. 8.20). Actually, the pulses should be regularly spaced and the pulse width at a particular position should be proportional to the area under the sine wave at that position. In the PWM

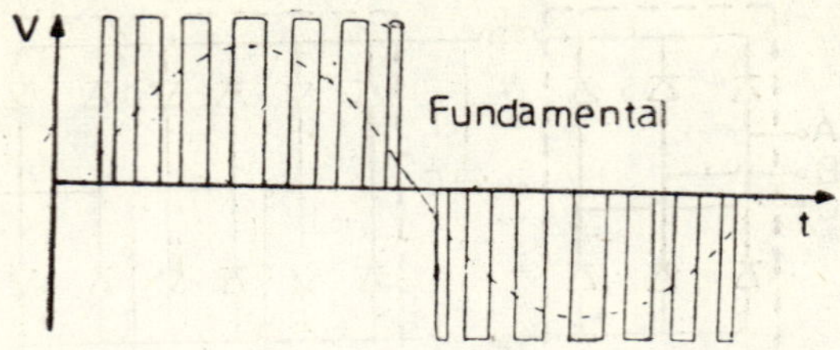

Fig. 8.20. Output voltage waveform of a PWM Inverter with sinusoidal modulation of the pulse width.

waveform, the lowest harmonic frequency is at the pulse repetition frequency and, if this is much higher than the fundamental frequency, adequate filtering is provided by the machine inductance. These waveforms are usually produced by means of a control circuit in which a high frequency triangular waveform is mixed with a sinusoidal waveform of the desired frequency. Voltage control is obtained by varying the widths of all pulses without affecting the sinusoidal relationship. A control voltage, which is derived from the reference oscillator, determines the output volts per hertz.

A three phase PWM waveform is obtained by using the bridge inverter of Fig. 8.16. The basic circuit is usually modified so that each thyristor has its own commutating circuit which can interrupt the flow of current at any desired instant. With the pulse width sinusoidally modulated, the output voltage and current have less harmonic content. The use of diode bridge rectifier and a high pulse repetition frequency in the inverter minimizes both distortion and interference of the mains.

While a high frequency reduces the lower order harmonic content and saves in filter cost, it linearly increases commutating losses in the inverter.

8.2.3 Cycloconverters

The cycloconverter is a 'direct' ac to ac frequency changer. The term 'direct' means that the energy does not appear in any other form than ac input or ac output contrary to an inverter which converts dc to ac.

While both the cycloconverter and inverter have as their output ac power of variable voltage and frequency, methods of attaining these outputs are quite different for the two circuits.

The single phase centre tapped circuit is shown in Fig. 8.21. This circuit is similar to a single phase centre tapped rectifier except that two groups of controlled rectifiers are provided to permit current reversal through the load.

Fig. 8.21. Single phase cycloconverter.

The thyristors P_1 and P_2 will provide positive current to the load and the thyristors N_1 and N_2 provide negative current. Let us assume that a firing circuit that can modulate the firing points of the controlled rectifiers in accordance with a control voltage, called the reference voltage, is present. If we use a sinusoidal reference voltage of a frequency that is low with respect to the line frequency and of a sufficient magnitude to vary the firing points of the controlled rectifiers from full retard to full advance, we will be able to reproduce a load voltage similar to the reference voltage (at least in average value) with some ripple occurring at twice the line frequency. In this operation, it is assumed that during the positive half of the reference voltage the P group of controlled rectifiers are delivering the current and during the negative half of the reference voltage, the N group of controlled rectifiers are delivering current to the load. The waveform of load voltage during this type of operation is illustrated in Fig. 8.22. From this waveform it can be seen that the low frequency output voltage is synthesized from the higher frequency line voltage. The description made above is a somewhat simplified picture of cycloconverter operation as it gives the voltage waveform produced by having no more than one rectifier in conduction at any particular instant of time. The load is assumed to be resistive although most of the loads encountered in practice will be a combination of resistance and inductance. The voltage waveform and the firing control with such loads will be slightly different and some of these differences will be explained below.

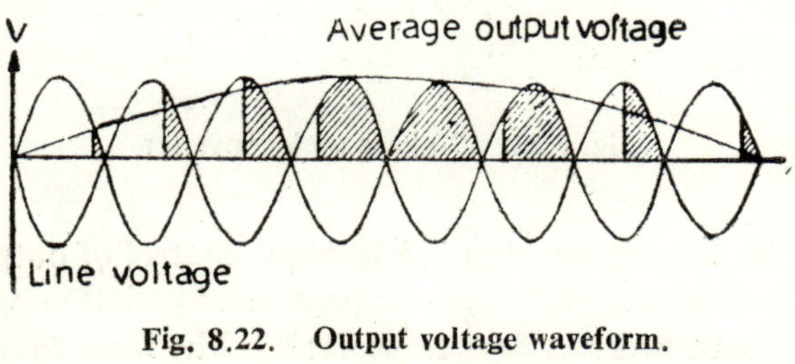

Fig. 8.22. Output voltage waveform.

It must be noted that either group (P or N) of controlled rectifiers mentioned in the above paragraph is capable of providing output voltage of either polarity, but that a given group is capable of conducting current in only one direction. With motor load, the current normally will be out of phase with the voltage. For example, if the load presents a lagging power factor, there will be intervals of time during which the current in a given phase is positive, although the average phase voltage is negative. To provide for this, both groups of thyristors must be supplied with appropriate firing signals so that both groups are producing the same average voltage, allowing current to flow in either direction. As the P group of thyristors is advanced in firing angle to produce more positive voltage, the N group firing angles are retarded to produce less negative voltage and therefore both groups are producing the same average load voltage.

Using the thyristor firing pattern described in the preceding paragraph, there will be intervals in time when point X_1 of the supply transformer (Fig. 8.21) is connected to point X_2 by virtue of thyristor P_1 and N_2 both being triggered. The purpose of the reactor shown in the same figure is to limit the magnitude of the current in this loop. In modern cycloconverters different methods are employed to suppress this circulating current between groups and thereby reduce the size of the reactors or eliminate them completely.

The basic circuit can be extended to polyphase operation by simply providing additional controlled rectifier elements in the positive and negative groups. The basic power circuit scheme of a three phase cycloconverter is shown in Fig. 8.23.

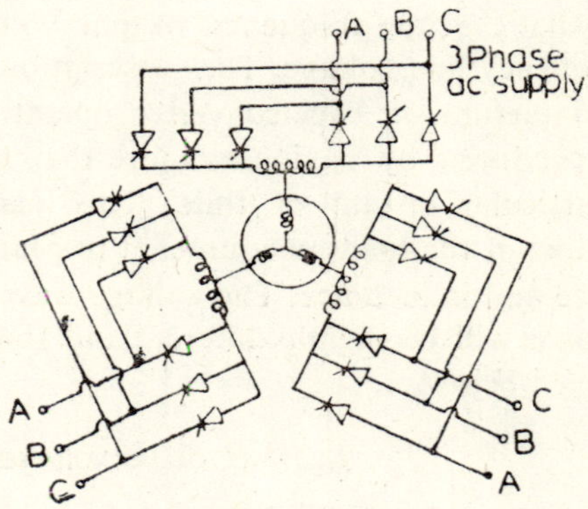

Fig. 8.23. 3-phase cycloconverter.

It should be pointed out that independent control of output frequency and voltage is obtained with only one parameter variation, viz., the firing points of the controlled rectifiers to be varied. The frequency of the output voltage is controlled by the rate at which the firing points are varied about the quiescent point and the output voltage is controlled by the maximum excursion of the firing points from the quiescent point. The cycloconverter, with its associated firing circuit, produces an output voltage that is a replica of the reference voltage.

The operation of a cycloconverter is characterized by several features. They are, generally, used as step down frequency converters. This follows from the fact that the low frequency ac voltage is synthesized from sections of the higher frequency input voltages. There is no fixed minimum ratio of input to output frequency; however, a 3 : 1 ratio is often found to be the limit for practical applications. Below this ratio, the efficiencies of both the cycloconverter and the motors fed by them start falling significantly.

Reversibility is another feature of cycloconverter drive systems. A cycloconverter fed ac motor drive will respond to a change in polarity of the input signals by changing direction of rotation of the motor without the use of contactors to reverse phase sequence.

The ability of the cycloconverter to handle power flow in either direction is another important feature. This, together with the abovementioned reversibility feature, provides an induction motor drive capable of operating in any of the four quadrants of the motor's speed torque curve.

While the cycloconverter has many attractive characteristics from a theoretical point of view, there are several limitations for its wide application in practice. It requires more power semiconductors than an inverter. For example, the 3-phase cycloconverter needs 18 thyristors, whereas the rectifier-inverter combination of Fig. 8.15 needs only 12 thyristors. Line pollution with harmonics and low power factor can also be problems with cycloconverters of high power rating.

8.2.4 Pulse Controlled Resistance

Conventional method of rotor resistance control of slipring induction motor demands simultaneous and precise variation of all the three balanced resistors in each phase. Often this is difficult to achieve. To overcome this limitation, a high frequency thyristor chopper, which enables the external resistance to be varied simultaneously and steplessly is employed.

Figure 8.24 shows a circuit in which the rotor slip power is rectified in a diode bridge rectifier and fed through a filtering choke to an external resistance. The thyristor in the chopper connected across the resistor is switched on and off at a high frequency. The ratio of on-time to off-time determines the effective value of rotor circuit resistance and thus controls the motor speed by changing its speed torque characteristics.

The most commonly used chopper circuit is shown in Fig. 8.25. TH_1 and TH_2 are the main and auxiliary thyristors; C is the commutating capacitor; diode D_2, C and inductor L form a part of the commutating resonance circuit; R_1 represents the external resistance to be varied and L_f, the

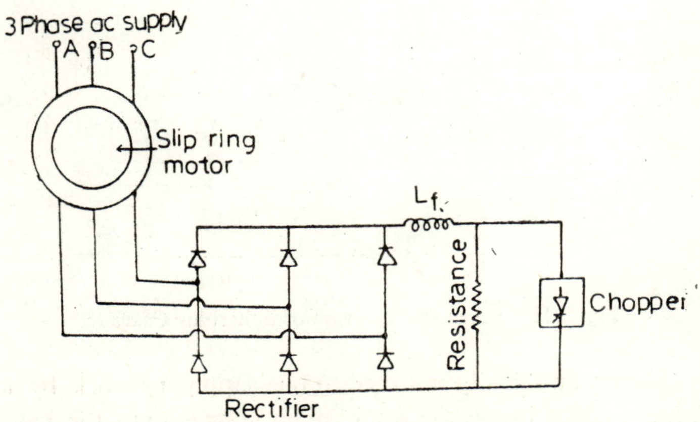

Fig. 8.24. Pulsed rotor resistance control.

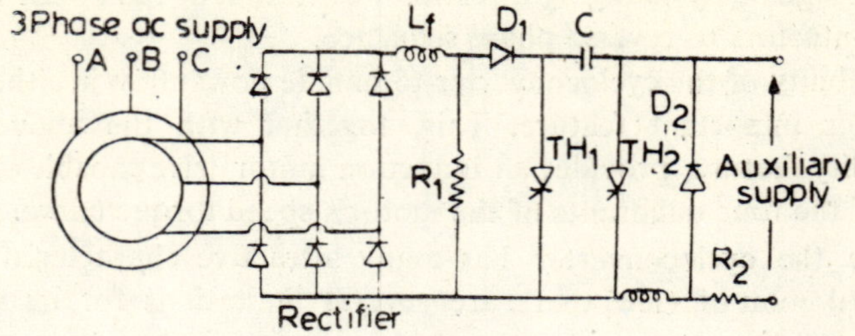

Fig. 8.25. Basic chopper circuit used for pulsed resistance control.

filtering inductance. An auxiliary voltage source V is required to ensure that commutation takes place at all speeds of the motor. A finite resistance R_2 has to be introduced to limit the power drain from the auxiliary source.

8.2.5 *Static Scherbius Cascade*

The schematic diagram of the Scherbius cascade with static converters is shown in Fig. 8.26. The induction motor is started using three rheostats in the rotor circuit. The ac slip power is first rectified by the three-phase diode bridge, then turned back into ac power at line frequency by the

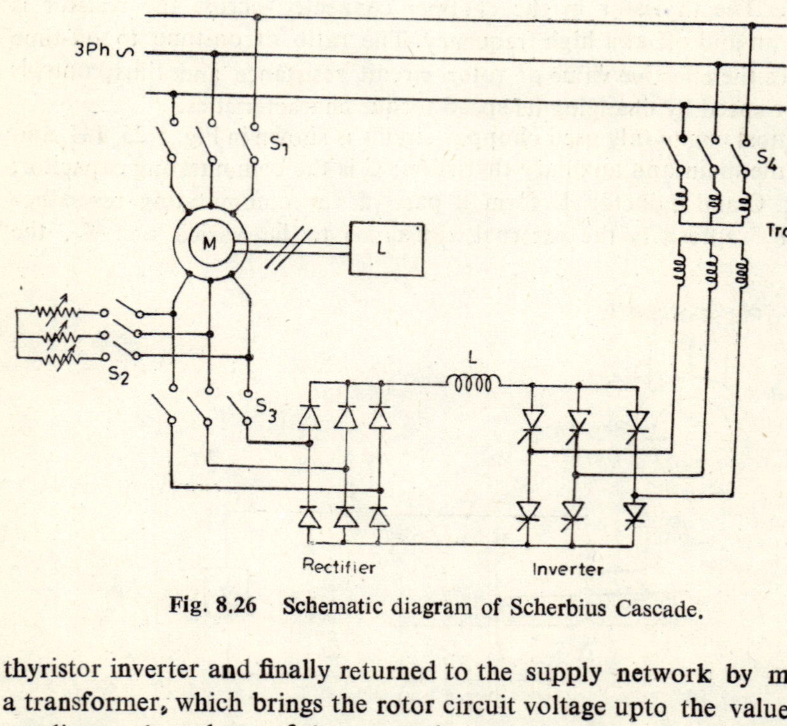

Fig. 8.26 Schematic diagram of Scherbius Cascade.

thyristor inverter and finally returned to the supply network by means of a transformer, which brings the rotor circuit voltage upto the value corresponding to the voltage of the ac supply network. The speed of the induction motor is regulated by the controlling the firing angle of the inverter. The gate pulses are provided by the firing circuits, synchronized with the supply voltage. Both the rectifier and the inverter are line-commutated by the alternating emfs appearing at the sliprings and supply network respectively. The average counter emf of the inverter may be considered as an injected emf opposing the rectified rotor voltage.

The system is started by switching on first S_1 and then S_2 while switches S_3 and S_4 remain off. As soon as the motor attains a steady speed, the rectifier-inverter combination as well as the transformer is connected to the supply network by switching S_2 off and S_3 and S_4 on.

8.2.6 Static Kramer Cascade

Figure 8.27 shows the schematic diagram of the Kramer cascade with static converter. A slipring induction motor's rotor circuit feeds the slip

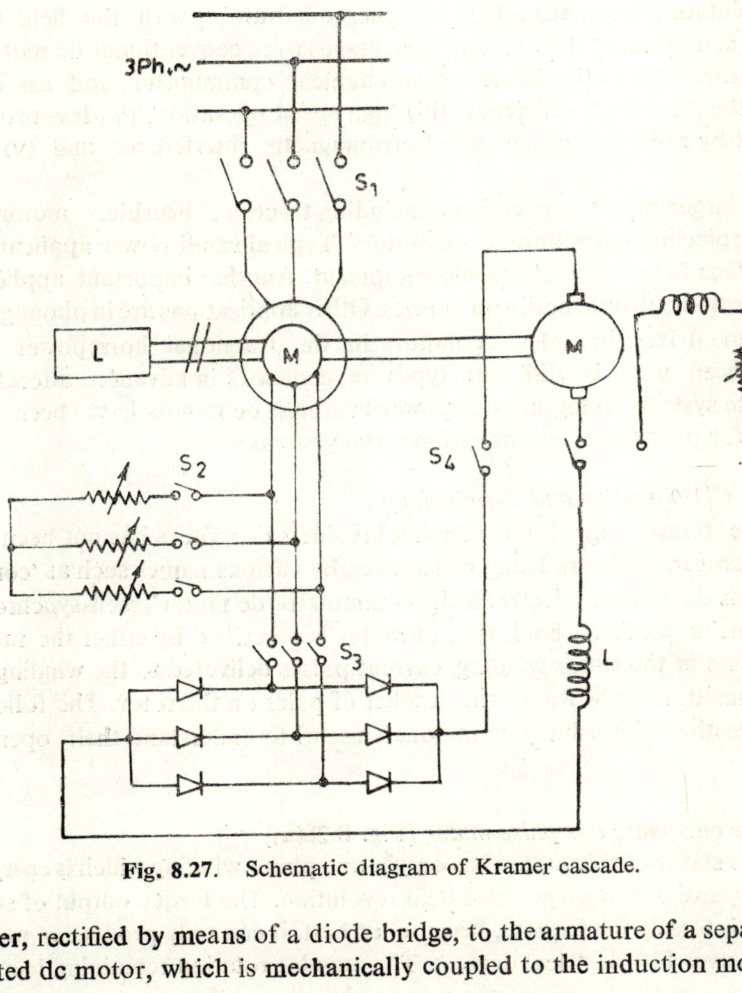

Fig. 8.27. Schematic diagram of Kramer cascade.

power, rectified by means of a diode bridge, to the armature of a separately excited dc motor, which is mechanically coupled to the induction motor.

The system is started by switching on S_1 first and then S_2, while switches S_3 and S_4 are off. As soon as the motor attains steady speed, the dc motor is energized by switching S_2 off and S_3 and S_4 on. Speed control is achieved by varying the field current of the motor. An emf proportional to the back emf of the dc motor may be considered to be injected into the rotor circuit of the induction motor to cause variation in speed of the system.

8.3 Brushless dc Motors

Unlike a conventional dc motor, a brushless dc motor has an "inside-out" construction, that is, the field poles rotate and the armature is stationary. The field poles consist of permanent magnets mounted on the inside of a steel cylinder and the armature is wound on a slotted laminated iron structure. The armature coils are switched by transistors or silicon controlled rectifiers (instead of the commutator) at the correct rotor position to maintain the armature field in space quadrature with the field poles. This arrangement has several advantages over conventional dc motors or ac motors such as (i) absence of mechanical commutator and associated problems, (ii) high efficiency, (iii) high speed operation, (iv) less problems caused by radio frequency and electromagnetic interference and (v) long life.

In larger power applications, including traction, brushless motors are fast replacing conventional dc motors. Typical small power application is in dc fans for cooling electronic equipment. Another important application is for spindle drives for disc memories. Other applications are in phonographs and tape drives. Brushless dc motors in the fractional horsepower range have been used in different types of actuators in advanced aircraft and satellite systems. Integral horsepower brushless dc motors have been developed for propulsion and precision servo systems.

8.3.1 Classification and Terminology

The terminology for describing brushless dc motors has not been standardized yet. They are being called even by various names such as 'commutatorless dc motor', 'electronically commutated dc motor', 'self-synchronous machine' and others. Each type of motor is described by either the number of phases of the stator winding, current pulses delivered to the windings by the transistors or SCRs, or the number of poles on the rotor. The following classification of brushless dc motors is useful to understand their operation too:

(1) *The one-phase one-pulse motor* [Fig. 8.28(a)]

The stator of this motor has only a one-phase winding which is energized by a transistor once per electrical revolution. The torque output of such a motor is totally inadequate, because, at best, it can only produce a positive torque over 180 electrical degrees. The angular rotation remaining has to be overcome by the inertia of the rotor or by means of auxiliary torques.

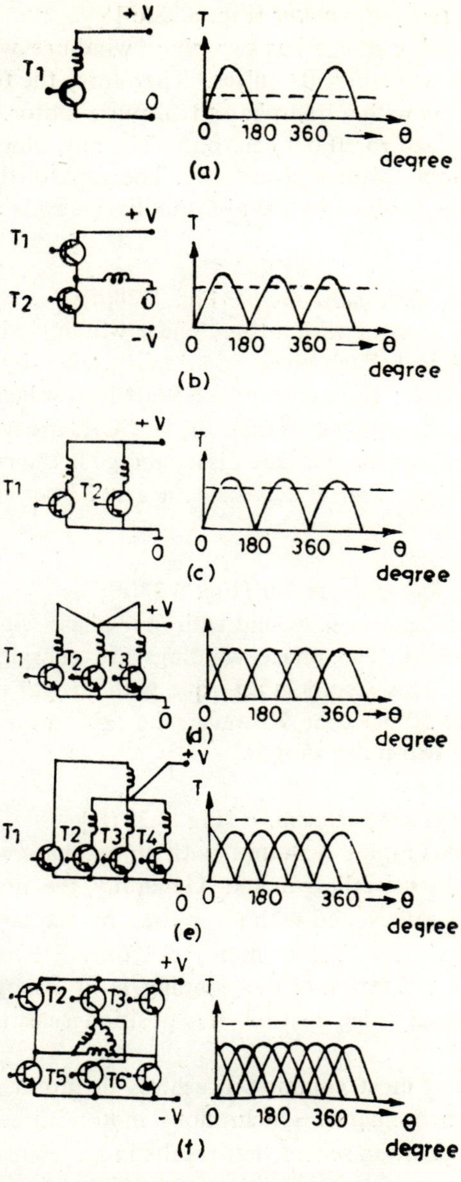

Fig. 8.28. Different types of brushless dc motors.

(2) *The one-phase, two-pulse motor* [Fig. 8.28(b)]

The stator of this motor has also a one-phase winding only but receives two pulses, that is, its winding is energized by two current pulses of opposite directions. The resulting torque distribution is, therefore, more favourable than with the one-pulse motor. Still a continuous electromagnetic torque is not achieved. There are still small regions without torque which have to be bridged with suitable auxiliary means. The advantage of this motor is its simple design, yielding a high utilization of the armature material.

(3) *The two-phase, two-pulse motor* [Fig. 8.28(c)]

The stator of such a motor has two phase windings which alternatively are energized by two current pulses. Therefore, the torque generated is basically the same as with a one-phase, two-pulse motor. Nevertheless, the winding will be utilized to 50 per cent only. The advantage of this motor is to be seen in its simple control electronics. The gaps of the electromagnetic torque have to be bridged by suitable auxiliary means as with one-phase motor.

(4) *The three-phase, three-pulse motor* [Fig. 8.28(d)]

This motor has a stator with a three phase winding which is displaced in space by 120° electrical. Each phase winding is excited by one pulse, that is, per electrical revolution; three current pulses are fed cyclically to the stator. The fact that only three power transistors or SCRs are required is the main advantage of this motor design. One disadvantage is the relatively low utilization of the winding (on the average, nearly 33 per cent) as well as the necessity of three position sensors.

(5) *The four-phase, four-pulse motor* [Fig. 8.28(e)]

The stator of this motor is wound with four-phase windings displaced in space by 90° electrical. The phase windings are energized cyclically with four current pulses. This results in a torque without gaps and a utilization of the winding upto 50 per cent. However, the expenditure for the electronics is twice that of the two-pulse design.

(6) *The three-phase, six-pulse motor* [Fig. 8.28 (f)]

The stator of this motor is wound with three-phase windings which can be connected in delta as well as in star. Generally, the neutral point is not used. The windings are excited with six pulses by six power transistors or SCRs in cyclic sequence. Such a motor not only delivers an even torque output but also the utilization of the winding is at its optimum. Its disadvantage is the relatively high cost for the position sensors and the control electronics.

This leads to the most common brushless dc motor—a combination of three-phase permanent magnet synchronous motor, three-phase solid state inverter and rotor position sensor that results in a system producing a linear speed torque characteristic as in the conventional permanent magnet dc motor.

8.3.2 Three-phase, six-pulse Brushless dc Motor Configuration

Figure 8.29 illustrates the schematic representation of 3-phase, 6-pulse brushless dc motor using a transistor inverter as the dc to ac converter. Where high power requirements exist, SCRs are used instead of transistors. Other trade off features like component cost, component reliability and simplicity of the inverter circuitry (relating to the need of commutation circuitry to turn

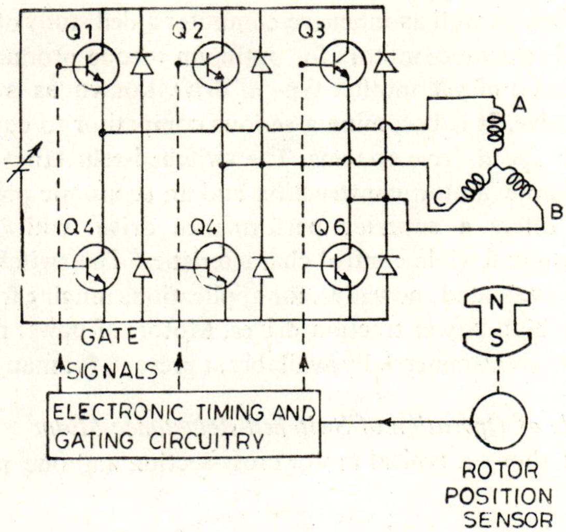

Fig. 8.29. Schematic of 3-phase, 6-pulse brushless dc motor.

off SCRs, a feature unnecessary for transistors) are of importance where SCRs and transistors of comparable power handling capabilities are available.

An integral part of the brushless dc motor system is the rotor position sensor. Although several methods are available for sensing angular position, the most commonly used ones are Hall effect sensors and electro-optical sensors.

Figure 8.30 shows typical performance characteristics of a 3-phase, 6-pulse brushless dc motor.

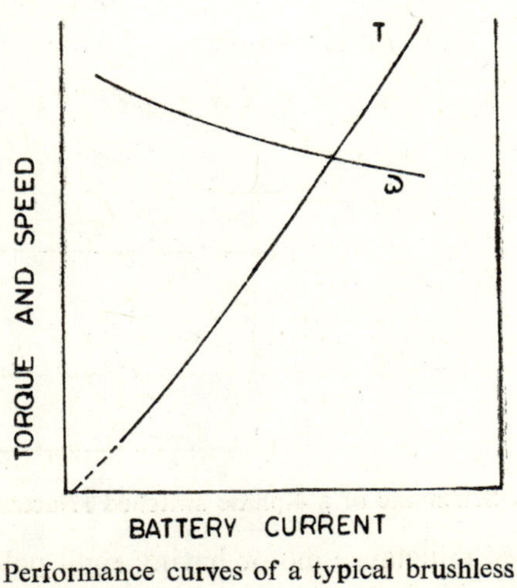

Fig. 8.30. Performance curves of a typical brushless dc motor.

8.4 Switched-Reluctance Motor Drives

The increasing availability and progressive improvement of power semiconductor devices as well as intensive computer aided study of the geometry of conventional reluctance motors for optimum torque production have led to the introduction of yet another type of drive konwn as switched-reluctance motor drive. It is becoming a serious competitor to converter fed dc and ac variable speed drive systems. The switched-reluctance drive system combines a simple motor construction and an economic power converter circuit which offers a superior performance drive with high overall efficiency and quite flexible control characteristics. The switched-reluctance motor is being evaluated, nowadys, for applications ranging from low power servomotors to high power traction drives. Motors of power ratings varying from 4 to 22 KW are commercially available at present, for many applications.

8.4.1 Principle of Operation of Switched-Reluctance Motor

Figure 8.31 shows a typical motor cross-section and one phase winding

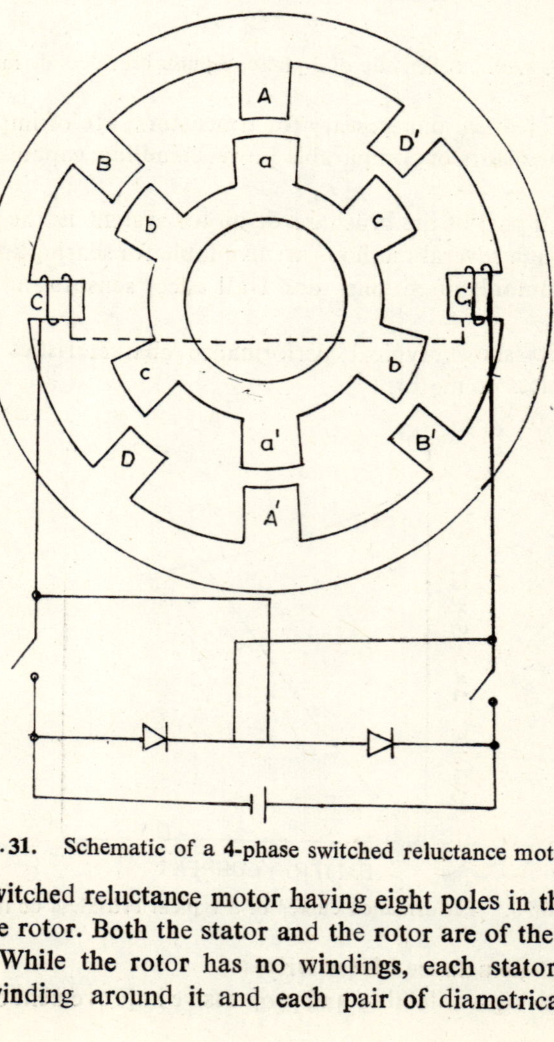

Fig. 8.31. Schematic of a 4-phase switched reluctance motor.

of a 4-phase switched reluctance motor having eight poles in the stator and six poles on the rotor. Both the stator and the rotor are of the salient-pole construction. While the rotor has no windings, each stator pole has a concentrated winding around it and each pair of diametrically opposite coils comprise one phase of the motor. Torque is produced by exciting any phase of the stator winding by means of unidirectional currents, this results in the magnetic attraction of an adjacent rotor pole as it tends to align into a position of minimum reluctance. When the number of stator and rotor poles differ, the sequential switching of the excitation from one set of stator poles to the next, in synchronism with the rotor position, results in an almost constant torque causing uniform rotation. The synchronization of the switching on of the excitation with rotor position can be accomplished with simple rotor position feedback.

8.4.2 Nature of Torque Production

Neglecting non-linearity of the magnetic circuit, the instantaneous torque developed in such machines can be expressed as

$$T = \frac{1}{2} i^2 \frac{dL}{d\theta} \tag{8.1}$$

where i is the instantaneous current in the exciting winding and L, the self inductance of that winding varying as a function of the angular position of the rotor.

It may be noted that the torque created is independent of the direction of current flow in the windings, so that unidirectional currents can be used permitting simplification of driving power circuit configurations.

Figure 8.32 (a) and (b) illustrate the ideal variation of inductance of the exciting winding with respect to the angular position of the rotor over a

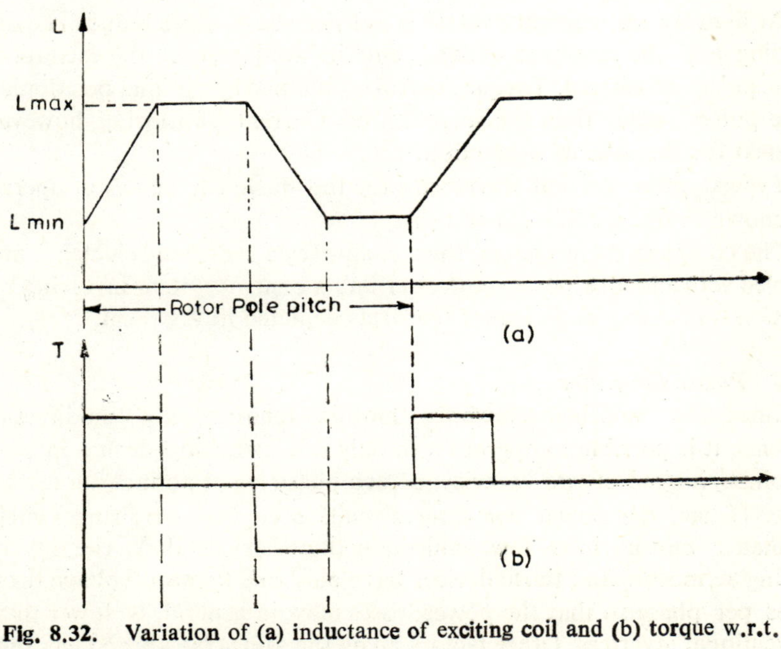

Fig. 8.32. Variation of (a) inductance of exciting coil and (b) torque w.r.t. rotor position.

periphery of one rotor pole pitch and the corresponding torque obtained, for an assumed value of constant current (using Eqn. 8.1). respectively. It can be seen that the torque can be controlled by appropriate switching on and off of the exciting current during the cycle of variation of inductance. An average motoring torque is produced by switching on current pulses in each stator phase to coincide predominantly with the period during the inductance value is increasing. Thus, a rotor position sensor is a must to determine the switching instants for different phases.

If the switching is delayed such that the current pulse period coincides with the period at which the rate of change of inductance with position is negative, torque produced will be negative and regeneration takes place. It may be worth noting that regeneration is achieved without any additional circuit elements. In fact, the possibility of operating in all the four quadrants of the speed-torque plane and obtaining flexible speed-torque characteristics simply by appropriate switching of current pulses makes the motor very versatile.

8.4.3 Modes of Operation of the Motor

Two distinct modes operation exist, corresponding to low or high speed. It is essential to monitor the exciting current during low speed operation, since each phase period is of long duration and the energization must be 'chopped' to restrict each phase current within the semiconductor ratings. Moreover, control over torque produced is achieved by varying the mean phase current and, hence, accurate monitoring of the current is required to obtain the high degree of controllability possible.

At high speeds, current control is not necessary, since inductance of the winding and the motional counter emf induced restrict the excitation to single pulses of current. Torque control is obtained by optimal positioning of these pulses rather than the current level. Current monitoring, however, is retained for the sake of protection.

Typical phase current waveforms for the above two modes of operation are shown in figures 8.33 (a) and (b).

The complete drive system, thus, comprises a switched-reluctance motor coupled with a load, a power converter and a control system involving rotor position transducer and current sensor as depicted in Fig. 8.34.

8.4.4 Power Converter

Since the switched-reluctance motors require only unidirectional currents, it is possible to operate with only one switching device in service per phase, instead of two in series in each phase leg of an inverter for an ac drive. Hence, the power converter circuits used for energizing switched-reluctance motors have few semiconducting devices than the inverters feeding ac motors and those devices have only one forward voltage drop in series per phase so that the power losses may, in general, be lower than in conventional inverters. Other factors being the same, the above facts should

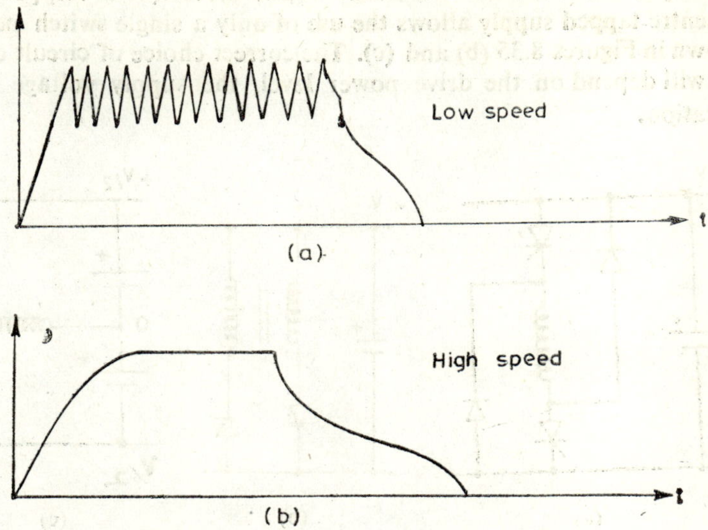

Fig. 8.33. Current waveforms at (a) low speed and (b) high speed modes.

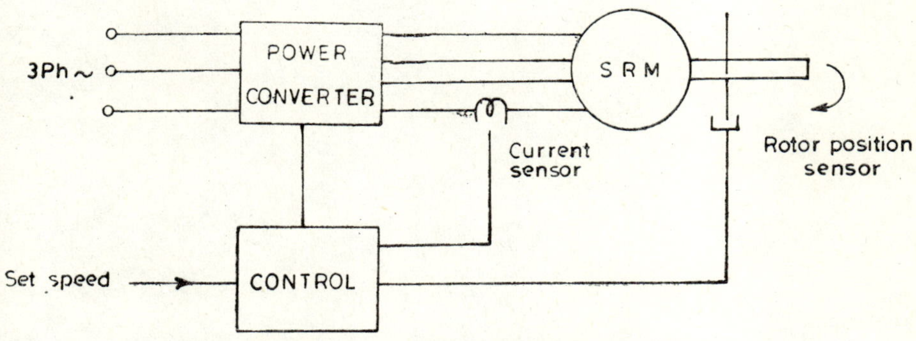

Fig. 8.34 Block diagram of a SRM drive system.

permit a reduction in the physical size of the converter and an increase in its reliability.

Also, shoot-through faults are impossible because there is always a motor winding in series with each main power switching device.

Different circuit configurations are available for the power converter for SRM drive, all of them having two essential elements—a controlled switch or switches to connect the direct voltage source to the exciting winding to build up current and an alternative path for the current to take when the switch is turned off. The latter is often provided by a diode or diodes such that the winding experiences a reverse voltage to collapse the current. Figure 8.35 shows three alternatives for one phase of a SRM power converter. A flexible circuit using two swithces per phase is illustrated in

Fig. 8.35(a), whilst the use of a bifilar wound motor (as in stepper motors) or a centre tapped supply allows the use of only a single switch per phase as shown in Figures 8.35 (b) and (c). The correct choice of circuit configuration will depend on the drive power level, the supply voltage and the application.

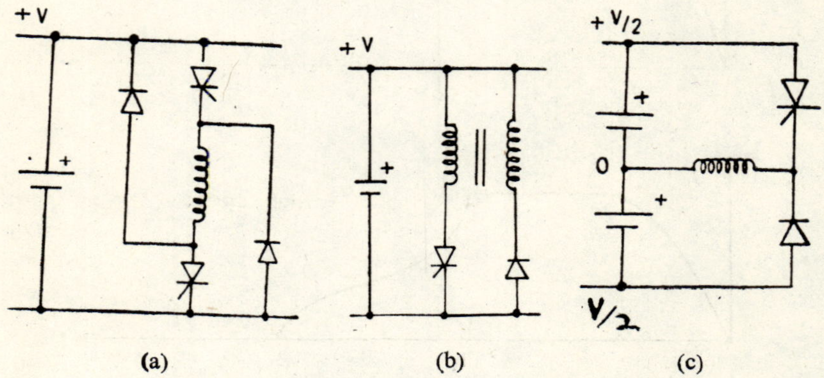

Fig. 8.35. Different power circuit configurations.

Chapter 9

Stepper Motors

Stepper motors convert electrical pulses received by their excitation (control) windings into discrete angular displacements—commonly called as steps. In fact, they function as decoders, transforming the digital information received in the form of electrical pulses into steps of angular position. Since there is a direct correspondence between the number of input pulses and the number of discrete angular steps taken by the motor, there is no need for a feedback transducer for measuring the angular position of the rotor. This means that automatic control systems can be built without feedback using stepper motors as actuators.

Stepper motors are widely used in numerically controlled (NC) machine tool positioning systems, since (i) the command signal in NC systems is in the form of numerical code or pulses derived from magnetic tape or punched paper and a stepper motor can operate directly off such command signals without requiring expensive analogue/digital (A/D) and digital analogue (D/A) converters; (ii) there is no need for a feedback transducer, which is quite cumbersome to instal on machine tools and (iii) the open loop system does not pose stability problems, thereby, considerably simplifying the system design.

Apart from NC machine tools, many stepper motor applications require periods of continuous motion at high speeds as well as occasional step-by-step rotation Examples include floppy disc drives, line printers, plotters and paper feeders.

Stepper motors are often constructed with a multipole, multiphase stator winding similar to the conventional electrical machines. They usually have three or four phase windings wound for a number of poles determined by the desired angular displacement per input pulse. The rotors are either of the permanent magnet type or the variable reluctance type. Stepper motors are actuated by means of an external drive logic circuit. As a series of pulses of voltage is applied to the input of the drive circuit, it feeds suitable currents to the stator windings of the motor to make the axis of the magnetic field step around in synchronism with the input pulses. Depending on the pulse rate and the load torque including that of inertia, the rotor follows the axis of the magnetic field due to the torque produced due to the interaction of the magnetic field caused by the permanent magnet rotor or the variation of reluctance with respect to angular position of the rotor.

9.1 Stepper Motors with Permanent Magnet Rotors

The rotor made out of permanent magnet material is either of the salient pole or cylindrical type. The stator has a two or three or four phase winding located in a slotted structure. The number of slots per pole per phase is usually chosen as one in multipolar machines.

Figure 9.1 shows a schematic representation of a 2-pole, 2-phase permanent magnet stepper motor, capable of making discrete steps of 90° as soon as voltage pulses are applied to the two phases of the exciting winding in a specified sequence. The axis of the magnetic field can have four different

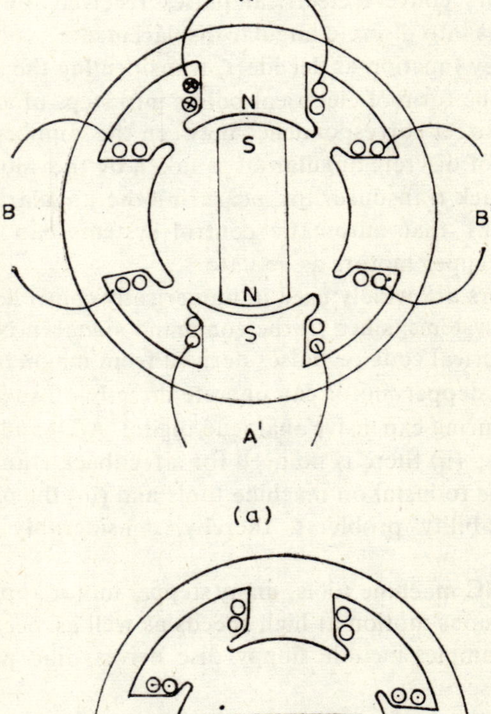

(a)

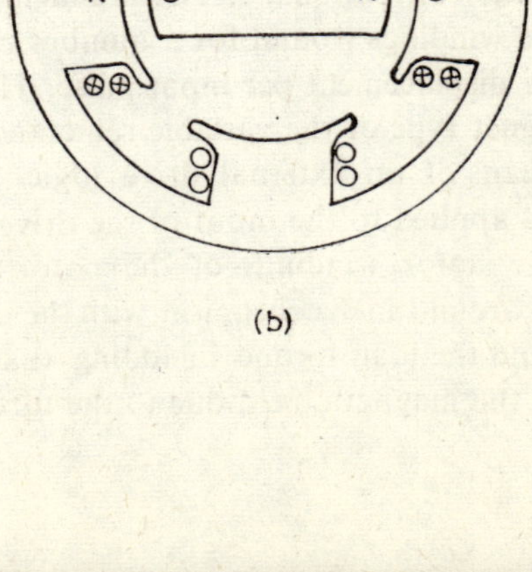

(b)

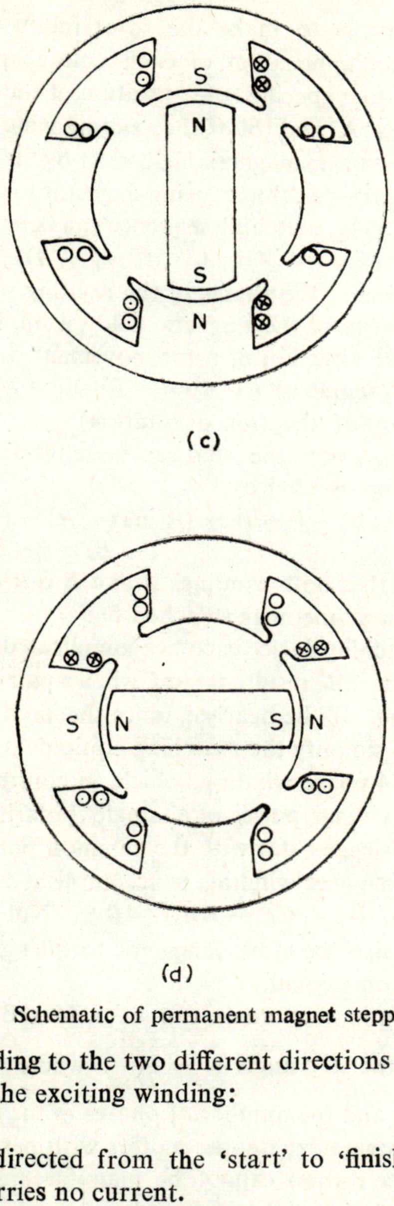

Fig. 9.1 Schematic of permanent magnet stepper motor.

positions corresponding to the two different directions of flow of current in phases A and B of the exciting winding:

(i) Current is directed from the 'start' to 'finish' of Phase 'A' while Phase 'B' carries no current.
(ii) Current flows from the 'start' to 'finish' of Phase 'B', while Phase 'A' doesn't carry any current.
(iii) Current is directed in Phase 'A' from 'finish' to 'start', with no current in Phase 'B', and
(iv) Current passes through Phase 'B' from 'finish' to 'start', with no current in Phase 'A'.

As a result of interaction between the magnetic fields caused by the exciting winding and the permanent magnet, electromagnetic torque is produced in such a way as to make the rotor follow the axis of the stator magnetic field. Hence, application of each voltage pulse to the exciting winding corresponds to a specific rotor position of the motor. The sequence of switching of phases 'A' and 'B' of the exciting winding mentioned earlier makes the axis of the stator magnetic field shift by 90° at every switching, as shown in Fig. 9.1 (a) —(d), thus causing the rotor to make discrete angular displacements of 90°. The switching sequence can be schematically represented as $(+A)-(+B)-(-A)-(-B)-(+A)$.... If the direction of current flow in any one of the phases of the exciting winding is reversed so as to reverse the polarity of the magnetic field set up, keeping the sequence of switching same, the direction of rotor movement would be reversed. (For example the switching sequence $(+A)-(-B)-(-A)-(+B)-(+A)$... would cause a reversal of direction of rotation).

It can be observed that the step can be reduced to 45° by adopting a switching sequence represented by

$$(+A)-(+A)(+B)-(+B)-(+B)(-A)-(-A)-(-A)(-B)$$
$$-(-B)-(-B)(+A)-(+A)...$$

$(+A)(+B)$ denotes that both windings A and B carrying currents in specified directions are simultaneously switched on.

Obviously the switching device becomes complicated and non-uniform torques are produced since the resultant mmf, when a pair of windings are simultaneously switched on, will be nearly $\frac{3}{2}$ times the mmf created by a single winding. In order to simplify the switching equipment, stepper motors are often wound with a 4-phase winding, which, in contrast to a 2-phase winding, can be fed with voltage pulses of a single polarity. The four phases are connected as a 4-legged star with the common point brought out. The switching of this four-phase winding, is accomplished in pairs in a sequence represented as $AB-BC-CD-DA-AB\ldots$ Not only the step angle remains as 45°, but also the electromagnetic torques produced during different switchings become equal.

Further reduction in step angle can be achieved by using multipolar machines. Actually the step angle $\alpha_s = \frac{360°}{2pm}$, where $2p$ and m represent the number of poles and the number of phases in the exciting winding. Due to difficulties in construction, stepper motors with permanent magnet rotors (even electromagnetic rotors) cannot be manufactured in small size with large number of poles and, hence, small steps are not possible. This disadvantage is overcome by the use of variable reluctance type stepper motors, since the number of poles in these are equal to the number of rotor teeth and the latter can be of a large number, as long as the tooth pitch is not less than 2mm.

9.2 Variable Reluctance Stepper Motors

The rotor is made out of slotted steel laminations and has no winding in it. The stator usually is wound for three phases. The stator windings are excited with the help of an external logic circuit in a specified sequence and the rotor seeks that position in which the reluctance between the stator and rotor is a minimum.

Figure 9.2 shows a schematic representation of a variable reluctance

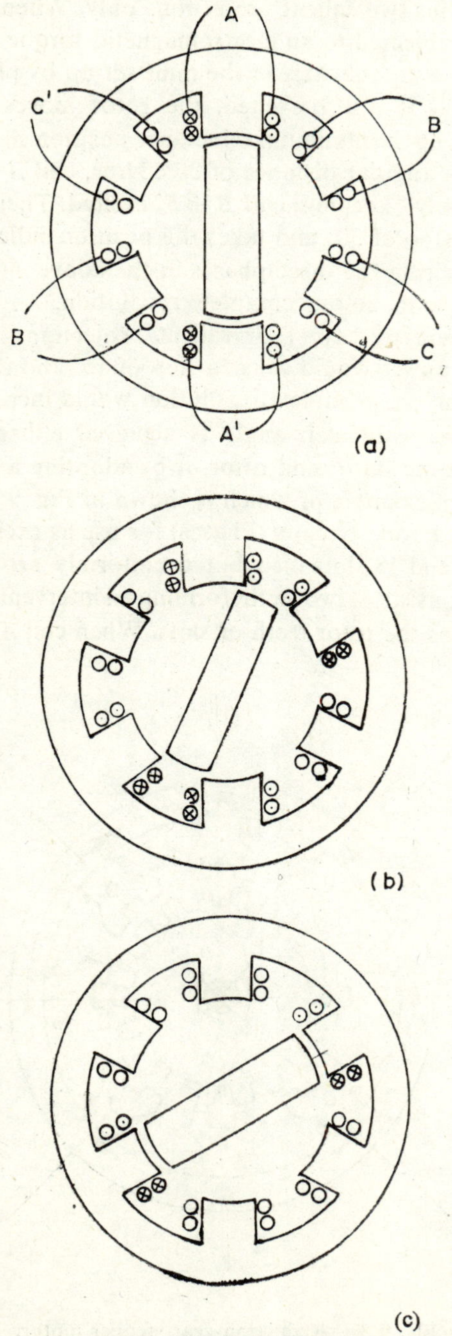

Fig. 9.2. Schematic of Variable-reluctance stepper motor.

stepper motor having six salient poles (teeth) with concentrated exciting windings around each one of them. The coils wound around diametrically opposite poles are connected in series and the three circuits (Phases), thus formed, are energized from a d.c. source through an electronic switching device. The rotor has two salient projections only. When coil $A - A'$ is excited the rotor is subjected to an electromagnetic torque and rotates until its axis coincides with the axis of the mmf set up by phase A (Fig. 9.2a). Next, when coil $B - B'$ is also excited, the rotor moves in the clockwise direction and takes up the minimum reluctance position shown in Fig. 9.2(b), thereby moving an angular distance of 30°. Now, coil $A - A'$ is disconnected from the supply, keeping coil $B - B'$ excited. Then, the rotor moves through a further step of 30° and takes the position indicated in Fig. 9.2(c). By successively exciting the three phases in the above manner, the motor takes twelve steps to make one complete revolution.

If the rotor were to have four salient projections instead of two, the successive voltage pulses would cause a step of 15° and the number of steps required for making one complete revolution would increase to 24.

A further decrease in step angle is achieved either by increasing the number of poles in the stator and rotor or by adopting a different constructional technique, an example of which is shown in Fig. 9.3. The stator has 8 salient poles and four circuits (Phases) for use as exciting winding. The rotor has 18 teeth and 18 slots distributed uniformly around. Each salient pole of the stator consists of two teeth, forming an intervening slot of the same angular periphery as the rotor teeth or slots. When coil $A—A'$ is excited, the

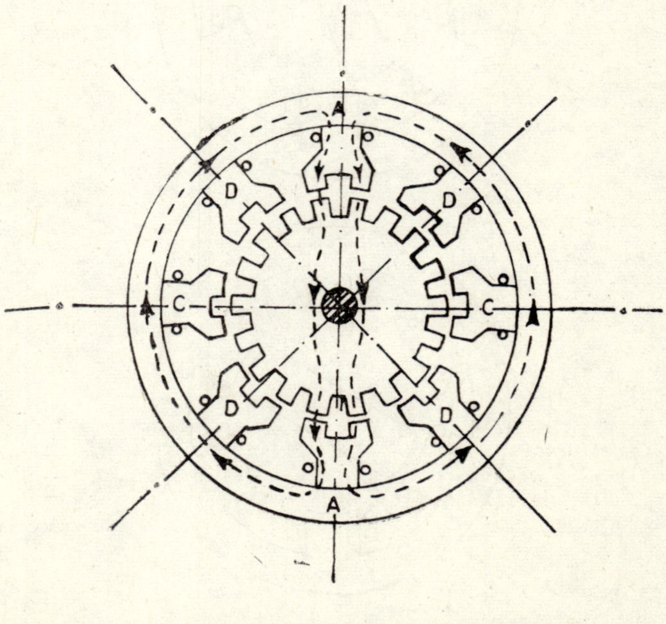

Fig. 9.3. Reduction-gear stepper motor.

resulting electromagnetic torque brings the rotor to the position shown in Fig. 9.3, where the magnetized teeth of the rotor and stator are exactly opposite to one another. When coil $B - B'$ is excited and the supply to coil $A - A'$ is disconnected, the electromagnetic torque moves the rotor clockwise by exactly one-half of the pitch of the rotor tooth, so as to bring two-rotor teeth opposite to the two magnetized stator teeth. The torque at once becomes zero, because of the resulting magnetic symmetry. Since the stator and the rotor teeth both have the same angular width, the step made is equal to 5°. By successive excitation of coils $A - A'$, $B - B'$, $C - C'$, $D - D'$, the motor makes 72 steps to complete one revolution. By choosing different combinations of number of rotor teeth and/or stator exciting coils, any desired step angle can be obtained; though the number of rotor teeth should not be as large as to make the tooth pitch less than 2 mm.

Since the type of construction, described above, enables us to achieve an electrical system of reduction gear, this motor is sometimes known as reduction gear stepper motor.

9.3 Stepper Motor Parameters

The most important parameters of stepper motors are as follows:

(i) Step angle, defined as the angular displacement of the rotor in response to each pulse;

(ii) Critical torque representing the maximum load torque at which the rotor does not move when an exciting winding is energized;

(iii) Limiting torque, defined for a given pulsing rate or stepping rate measured in pulses per second, as the maximum load torque at which the motor follows the control pulses without missing any step;

(iv) Synchronous stepping rate defined as the maximum rate at which the motor can step without missing steps. The motor can start, stop or reverse at this rate.

(v) Slewing rate, defined as the maximum rate at which the motor can step unidirectionally. The slewing rate is much higher than the synchronous stepping rate; but the motor will not be able to stop or reverse without missing steps at this rate.

9.4 Stepper Motor Characteristics

Torque-displacement characteristic gives the relation between the electromagnetic torque developed and the displacement angle θ of the rotor from its stable equilibrium position during steady state with one control winding permanently energized. This torque is obviously zero for the equilibrium position, i.e., at $\theta = 0$ with respect to the poles of the energized winding. An opposing electromagnetic torque appears as soon as an effort is made to move the rotor from this position. Initially, this torque increases with the angular displacement reaching a maximum value and, then, falls to zero. The nature of the variation of torque versus displacement is almost sinusoidal for motors with both permanent magnet rotor and variable reluctance rotor as shown in Fig. 9.4(a) and (b). Actually, they correspond to the

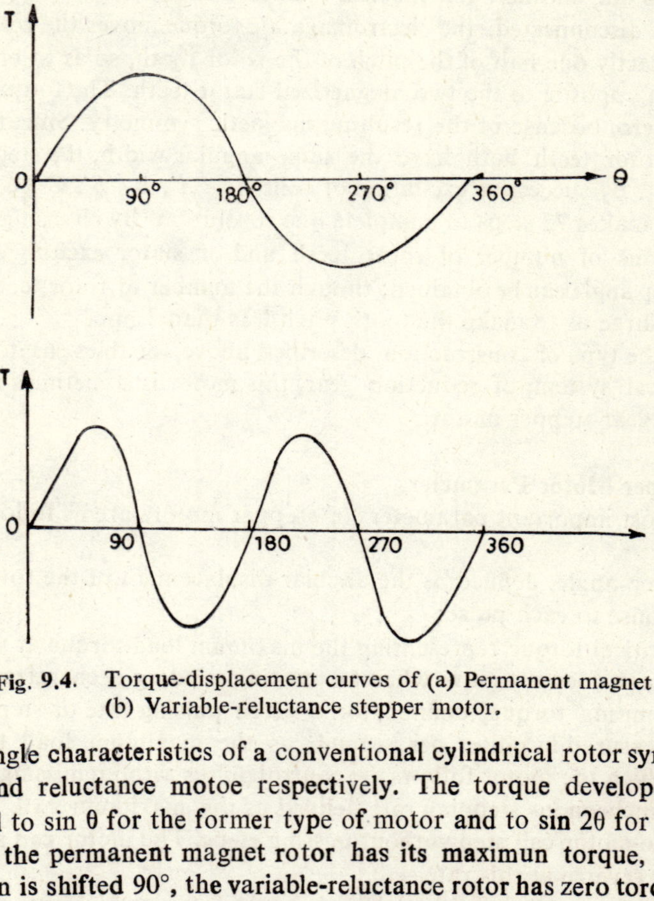

Fig. 9.4. Torque-displacement curves of (a) Permanent magnet (b) Variable-reluctance stepper motor.

torque-angle characteristics of a conventional cylindrical rotor synchronous motor and reluctance motoe respectively. The torque developed is proportional to sin θ for the former type of motor and to sin 2θ for the latter. Whereas the permanent magnet rotor has its maximun torque, when the excitation is shifted 90°, the variable-reluctance rotor has zero torque for 90° shift and can move in either direction. The permanent magnet rotor has the

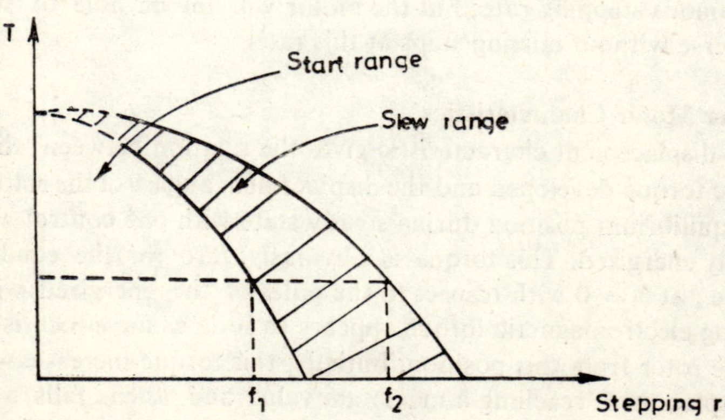

Fig. 9.5. Torque—stepping rate curves.

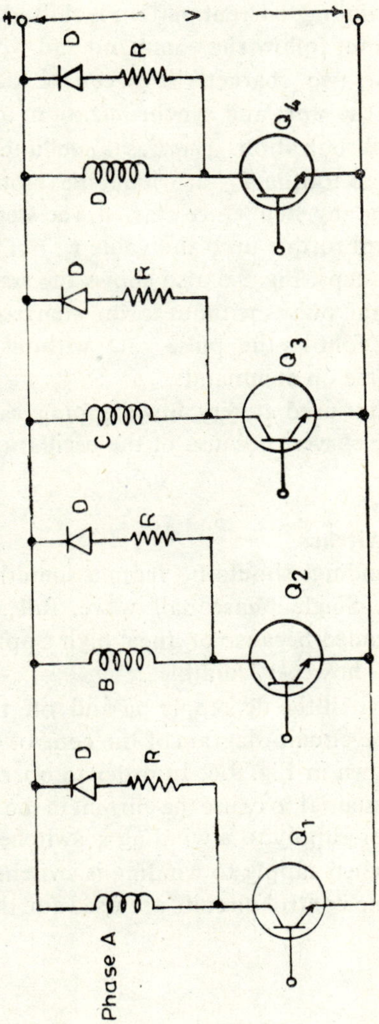

Fig. 9.6. Power converter for 4-phase stepper motor.

significant feature that the rotor position θ is defined by the winding currents without any ambiguity; the variable-reluctance rotor has two possible positions for each winding current pattern.

Torque versus pulsing rate characteristic describes the electromagnetic torque as a function of stepping rate in pulses per second. As the stepping rate is increased, the rotor has less time to drive the load from one position to the next as the stator winding current pattern is shifted. Beyond a certain pulsing rate the rotor cannot follow the command and would begin to miss pulses. For a given motor, two characteristics can be defined (Fig. 9.5)- curve 1 corresponding to the start and synchronization of the motor and curve 2 to the loss of synchronization. There is a definite stepping rate f_1 pulses /sec. for a given load torque T_L, at which the motor can start and synchronize without losing any step. Once started, the stepping rate can be increased for the same load torque upto the value f_2, but above this value the motor begins to lose steps. Fig. 9.5 also shows the start range in which the load position follows the pulses without losing steps and the slew range in which the load velocity follows the pulse rate without losing step, but cannot start, stop or reverse on command.

The stepper motor is not used at very low stepping rates (indicated by the dotted portions of the curves) because of the oscillations caused due to lack of damping.

9.5 Power Supply and Switches

Power fed to motor windings should be from a smooth dc supply with as little ripple as possible. Single phase half wave, full wave and bridge rectifiers are not recommended because of their high ripple content. Three phase bridge rectifiers are, however, suitable.

Power switches that switch the dc supply on and off the winding may use bipolar transistors. The circuit diagram of the control circuit for a four phase stepper motor is shown in Fig. 9.6. In order to operate the motor at high stepping rates, it is essential to cause the current in the winding to rise as fast as possible when power supply to a winding is switched on and to cause decay as fast as possible when supply to winding is switched off. Different types of rise and decay time control circuits are used for this purpose.

Chapter 10

Industrial Applications

10.1 Steel Mills

Rolling of steel, the process during which the cross section of the metal gets reduced, while increasing its length proportionately, is the major function of steel mills.

Steel mills, usually, produce blooms, slabs, rails, rods, sheets, strips, beams, bars or angles. Rolling mills are generally classified according to their products, since, very often, a particular mill is capable of manufacturing a specific product and the design of millstand, millhead, auxiliaries and drives changes with the type of product. A blooming mill produces blooms—any length of rolled metal having almost a square cross section. If the rolled metal is of rectangular cross section, it is called a slab and the mill that produces it is a slabbing mill. Blooms of reduced cross-section are known as billets. A billet can be further rolled into bars, squares, angles, rods etc. and the same mill can produce these varieties of different sizes and shapes. Such mills are termed as merchant mills. A plate rolling mill reduces slabs into plates, whose thickness is small compared to the width. Plates are changed to strips in a strip mill and to thin sheets in a sheet mill. In all the above mentioned types of steel mills, a range of sizes having similar shapes can be manufactured by changing the number of passes or by using stand rolls having the appropriate pattern.

Technologically the rolling mills are divided into four types: continuous cold rolling mills, reversing cold rolling mills, continuous hot rolling mills and reversing hot rolling mills.

Sheet steel having high mechanical properties, uniform gauge and good surface quality can be produced only by cold rolling. Blooms, slabs, billets, strips etc. are manufactured by hot rolling.

In a reversing mill there is only one stand carrying the rolls that press the metal and the metal is passed through this stand alternately forward and backward several times in order to reduce it to desired size. Each motion or travel is known as 'pass'. A continuous mill or tandem mill consists of several stands, each one of them carrying pressing rolls. The metal passes through all the stands in only one direction and gets rolled simultaneously by some or all of them.

Reversing cold rolling mills produce sheets with a minimum thickness of 0.15 mm and a maximum width of 2 m and more, black plates with a minimum thickness of 0.07 mm and a maximum width upto 1.3 m and bands with a minimum thickness upto 0.0015 mm and a maximum width upto 1 m. The turnover of a reversing cold rolling mill is limited by the low sheet speed owing to the continuous acceleration and retardation and the time taken for readjusting the roll gap during each pass. This disadvantage is overcome in continuous cold rolling mills, in which the metal passes in one direction, through several successive stands.

Continuous hot rolling mills, usually, manufacture billets, strips and the products of a merchant mill. Blooming and slabbing mills, which produce blooms and slabs respectively are generally, of reversing hot type.

10.1.1 Reversing Hot Rolling Mills

Process: Reversing hot rolling mills roll out blooms or slabs from hot steel ingots, which come directly either from steel making shops or from heating chambers called soaking pits. Red hot ingots are taken out one by one by a special crane having tongs and placed on a car known as ingot buggy. The ingot buggy travels a short distance on rails along the same line as the mill bed and takes the ingots to the front end of the mill bed. The mill bed consists of a large number of rolls.

Figure 10.1 shows a schematic diagram of the various stages in the hot reversing mills. The ingots brought upto the front edge of the mill bed are tilted and placed on the receiving table such that they lie with its longitudinal axis horizontally on the bed. Then, they move over to the weighing table, where it is possible to weigh the passing ingots. The travel of ingots over the roll tables is facilitated by the rotation of the rolls on which they are laid. From the weighing table, the ingots move over an approach table, several intermediate tables and a few front work tables before reaching the main working roll of the millstand. On the other side of the stand, the mill bed has a few back work tables, several intermediate tables and a

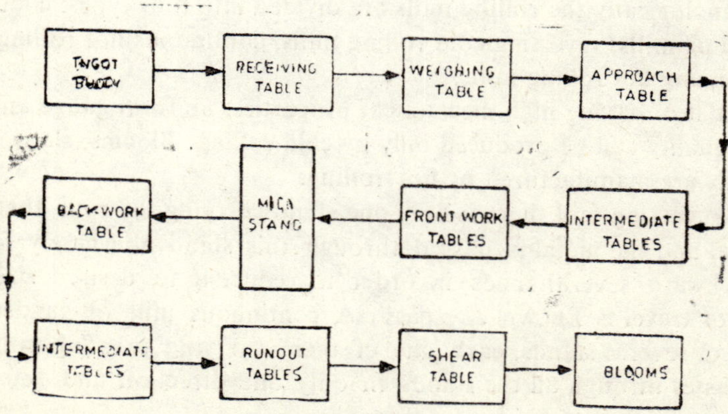

Fig. 10.1.

run out table in succession ending with a shear table, where the rolled metal is cut into standard lengths. The length of the mill bed on either side of the stand will be determined by the length of the product of the mill before shearing.

The ingots are rolled by the working rolls into blooms, as they pass through the mill stand several times in the forward and reverse directions. The gap length between the main rolls is reduced gradually in successive passes by bringing down the top roll of the main rolls by a screw down mechanism. This adjustment of gap length is made during the time when the forward ingot is stopped for reversal of its direction of motion.

Alignment of the metal to required entry positions in mill rolls is done by devices called manipulator side guides. Tilt fingers, in conjunction with manipulators turn metal over to permit rolling on all four sides.

Requirements of the drive: The main drive of a hot reversing mill consists of either a motor driving both rolls through pinions or a twin drive arrangement in which each roll is separately driven. A single armature motor may be used for mills of medium size, but a double armature motor is necessary for large mills. In very large mills a double armature twin drive may be used. Main requirements of the drive for the main working rolls are: (i) wide range of speed variation, (ii) ability of the motor and its control equipment to permit frequent starts and reversals, (iii) reversal of the direction of rotation of the motor without causing serious disturbances in the power handing circuits, (iv) automatic control of operation, (v) high reliability and (vi) accuracy.

In addition, there are a number of auxiliary drives which are used to convey the ingot, bloom or slab to or from the mill and in making mechanical adjustments to the mill or metal on a pass by pass basis. All these variable speed drives must be integrally controlled for effective functioning of the mill.

10.1.2 Continuous Hot Rolling Mills

Process: Continuous hot rolling mills, which produce billets or strips, have two groups of stands—roughing stand and finishing stand. Each stand may have two or four rolls, depending on which the stands are called two-high or four-high. In a four-high stand, the inner rolls of smaller diameter are the working rolls and the outer rolls are used to vary and maintain the gap between the working rolls. The metal passes through both sets of stands in only one direction. While the metal gets rolled simultaneously in the finishing stands, it need not be so in roughing stands, especially when the two groups of the latter are located far apart.

Requirements of the drive: In order to produce billets of different dimensions, different gap lengths between the working rolls are to be set. For a specific gap setting, the speeds of the stand motors are maintained at values differing slightly from each other to take care of the gradual reduction in thickness and lengthening of the metals, i.e., the speed of the motors must be capable of variation over a range of 1.5-2. In order to avoid sag of the metal between stands, the speed of the stand motors should be controlled to a high degree of accuracy. Also, the metal comes into contact with the working rolls at its speed of operation. Since this gives rise to a sudden application of load on the stand motor, its speed may drop and cause sag of the metal between stands. Hence, by means of an automatic control scheme, the speed of the motor must be restored to its set value as quickly as possible.

10.1.3 Reversing Cold Rolling Mills

Process: The strip to be rolled is received by the mill in the form of a reel, wound on a mandrel, mounted on one of the reel stands as shown in Fig. 10.2. Another stand carries an empty mandrel. The end of the strip is threaded manually through the stand rolls and round the empty mandrel. The mill is then run with a slow speed and the strip is wound on the coiling drum for a few turns with low tension and roll pressure. The speed is then increased with a uniform acceleration and the metal is rolled at the required pressure and tension. During this, one reel gets coiled and the other is uncoiled. When, only a few turns are left to be rolled, the speed is reduced and the direction is reversed, thus ensuring that the strip does not come out of the uncoiling reel. The same procedure is repeated while rolling in the opposite direction. A number of such reversed rolling with gradual gap length reduction produces the desired strip.

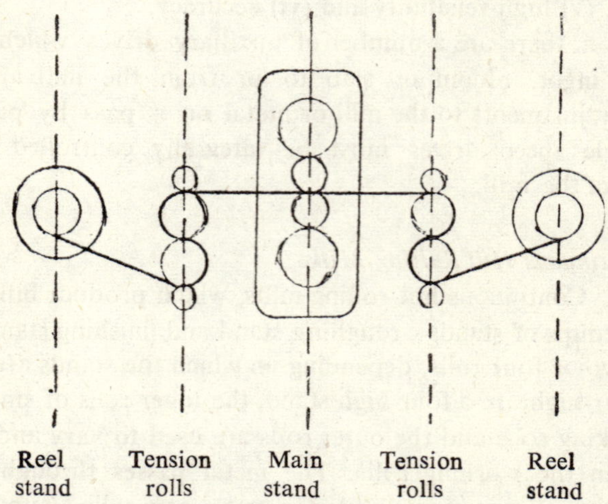

Fig. 10.2.

Requirements of the drive: The mill consists of a reversing stand, two reversible coilers and in most cases, an additional uncoiler for the first pass. The mill stand is, often, of the four high type, i.e., it has two work rolls with a diameter between 200 mm and 500 mm and two back up rolls, whose diameters are about two to four times greater than that of the work rolls. The back up rolls are required to prevent bending of the work rolls.

The torque required for rolling is provided by either one or two stand motors. The drive is provided to the back up rolls through gears and the work rolls are friction driven. Sometimes the work rolls are themselves driven by individual motors. The coiler and uncoiler are driven by separate motors. In addition to coiling and uncoiling the strip, the coilers also have the job of developing a defined strip tension between themselves and the stand during the individual passes. This tension is not only required for avoiding either looping or breaking of strips, but also to prevent variation in the gauge of the material being rolled. The roll gap is adjusted by means of a screw-down system.

One of the characteristics of a rolling mill is that the armature inertia of the mill motor would be larger than that of the rolls. Hence, in large capacity mills, to keep the ratio of motor inertia to load inertia low, motors with multiple armatures are chosen.

In principle, the mill stand requires a four quadrant drive at different speeds. Hence, the control system must provide both armature voltage control and field weakening. Field is kept at its rated value until the maximum back emf is reached and subsequently reduced depending on the speed requirement. When twin drives (separate motors for upper and lower rolls) are used, separate controls are provided for the two motors with additional circuitry to ensure equal load sharing.

The coilers and uncoilers have to maintain constant tension during various modes of operation of the mill. They should also be able to provide the necessary acceleration from threading to normal operating speed. The accelerating torque is dependent on the moment of inertia. In most of the cases, the inertia of the motor constitutes a considerable portion of the total inertia. Therefore, to reduce the inertia, it may be necessary to design the coiler motors with two or three armatures.

In order to maintain a constant tension in the strip, the armature current of the motor and the ratio of flux to effective diameter of the reel of the coil must be kept constant.

$$(F = K_1 \frac{IE}{V} = K_2 \frac{I\phi N}{DN} = \frac{K_2 I\phi}{D},$$

where F, I, E, V, ϕ, N and D denote, strip tension, armature current, induced emf, strip speed, field flux, speed of the motor and effective diameter of the reel respectively.) Since the effective diameter of the reel changes as the strip is unreeled at the input end and reeled up at the output end, the flux also is to be varied linearly with the coil diameter. If the maximum strip tension is required at maximum speed from the point of view of rating of the motor, the optimum condition is to make the field weakening range identical with the diameter range. However, this may not always be possible, since it would be very difficult to get a field weakening range better than 1:3. In such cases, it is necessary to operate the system both in armature and field control ranges, which demands a motor of higher power rating.

10.1.4 Continuous Cold Rolling Mills

The metal in such mills passes in only one direction through several successive stands. The stands are, often, of the four high type.

The mill consists of an uncoiler and coiler reel; the latter demands accurate torque and speed control. The most important features of continuous cold rolling mills are as given below:

The strips are subjected to a large interstand tension, which is kept at constant value depending on the desired thickness of the sheet to be produced. The threading of the metal through the stands is done at a low speed. As soon as the threading is over, the speed of the mill is raised to the working value. Again, it is brought to a low value, before the strip leaves the first stand. Depending on the products to be obtained, a wide variation in speed of the mill drive is called for.

10.1.5 Drives Used in Different Mills

Use of dc motors is almost universally accepted for choosing the main drives in both reversing and unidirectional continuous mills. Motors for reversing mills must have high starting torque, wide speed range, precise, speed control, be capable to withstand overload and pull out torque which may be as high as three times the rated value and have good commutation throughout. Acceleration from zero to base speed and then to top speed and subsequent reversal from top speed forward to top speed backward must be achieved in a couple of seconds. The moment of inertia of the armature must be as small as possible. Therefore, the motors used for steel mill duty are of totally enclosed, forced ventilated and higher class of insulation.

As regards controlling the speed of the drive, Ward-Leonard system with a flywheel is unsurpassed for the purpose. A liquid rheostat known as slip regulator is provided for the starting and regulation of the slip of the induction motor prime mover of the Ward-Leonard-Ilgner set. This is also used as a loading resistance for the induction motor while applying dc rheostatic braking to bring the Ilgner set to a quick stop at a fast controlled rate. DC supply required for rheostatic braking is obtained from the exciter of the variable voltage dc generator and the stored energy of the Ilgner set is dissipated in the slip regulator. In addition to this electrical braking, a mechanical braking arrangement is also provided to bring the Ilgner set to rest in case of emergency. Of course, nowadays, the Ward-Leonard-Ilgner system is rapidly losing its foothold against the thyristor converter, which has brought phenomenal changes in the design of control systems for the mill drives.

10.2 Paper Mills

Pulp making and paper making are the two main processes that take place in a paper mill and the drives required are quite different for each one of them.

10.2.1 Pulp Making

Pulp is made in two ways—purely by mechanical means and by both mechanical and chemical processing. The former involves grinding logs of wood of about a meter length on large grind-stones. Grinders operate at almost constant speed and can be started under light load conditions. Hence, synchronous motors are considered as most suitable for grinder drives. Since, they usually work with speeds of 200-300 rpm, geared drives are used, especially when the motors are installed in a separate room for protection from humid atmosphere. Usually, making pulp by purely mechanical means consumes more than fifty per cent of the total power requirement of a paper mill; hence, large size grinders driven by 3000-4000 kW motors are, normally, considered as economical.

Pulp can also be made by cutting logs of wood into chips of several centimetres length and treating them with alkalies along with other raw materials like grass, rags etc. During the chemical treatment the material is continuously beaten. Wood choppers have random load characteristics and their inertia is large, depending upon the size of the disc, on which the knives of the chopper are mounted. Beaters, usually are required to start with large load.

The end products of the grinders as well as beaters are refined and stored in large tanks as pulp ready for making paper.

Depending upon the size of a mill, the ratings of motor used for chipping, beating, refining and storing range from several hundreds to thousand kilowatts. Except for beaters, synchronous motors are used in these drives. Since beaters, very often, require speeds less than 200 rpm and large starting torque, slip-ring induction motor drives are found to be more suitable.

10.2.2 Paper Making

The machine that makes paper in a paper mill has to perform the job of forming sheets, removing water from sheets, drying of sheets, pressing of sheets and reeling up of sheets. Therefore, the paper is made in the following five sections: (i) Couch section (wire section), (ii) Press section, (iii) Dryer section, (iv) Calender section, and (v) Reel section.

Figure 10.3 shows a schematic layout of the different sections that make paper. The paper pulp suspension with a moisture content of 98 per cent to 99 per cent is transferred uniformly to the wire. Most of the quantity of water drips through the wire mesh and the rest is removed by suction. At the end of the wire section, the moisture content would have reduced to about 80 per cent. In the press section, which follows the wire section, sheet is pressed between woollen felts so as to squeeze out water from the wet sheet and the web leaves the press section with a moisture content of 65 to 60 per cent. In dryer section, the sheet is further dried by passing it over and under the heated cylinders until the desired dryness, usually 6 per cent of moisture content, is obtained. In calender section, sheet is subjected

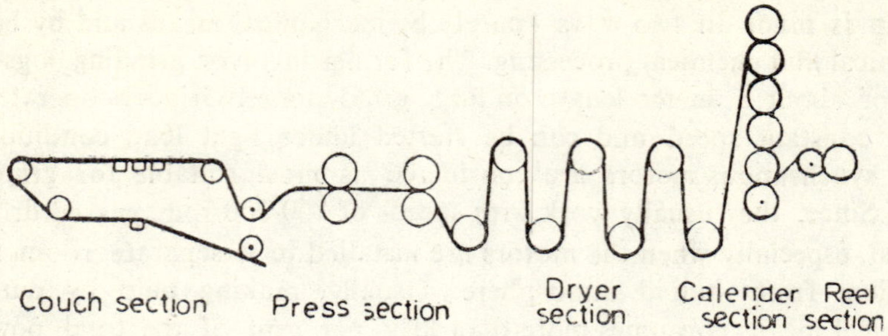

Fig. 10.3.

to pressure and friction so that a compact and smooth surface of sheet results. In the reel section, the sheet is wound up on a mandrel.

10.2.3 Requirements of Paper Machine Drive

1. For the actual formation and production of sheets and from the point of view of economy, it is necessary to maintain the speed of paper machine constant.

2. For a paper machine to be multipurpose, its speed should be adjustable over a range as large as 10 : 1.

3. In the wet end of the paper machine, the web becomes elongated by about 5 per cent. Care should be taken to allow for this elongation by varying the speeds of individual sections. As this elongation is a function of the quality of the paper, speeds of sections must be independently adjustable.

In dry end of paper machine, the web hangs freely between sections without any support. Therefore, a definite amount of tension must be set and regulated. This also necessitates a definite difference between the speeds of successive sections.

The relative speeds of two sections have a direct bearing on the pull on the sheet. Hence, steady state accuracy of 0.1 per cent or better is specified for the speed control system so as to avoid tearing and folding of sheet.

4. As web is manually introduced into the calender, an arrangement to take up sag is a must.

5. In the last two sections, viz., the calender and the reel, variations in tension in sheet can occur even if the correct relative speeds are maintained due to uneven drying and other factors. It is, therefore, imperative to augment the speed control circuit by an overriding tension control.

6. While cleaning the wire, it has to be moved forward a few centimetres at a time so that it can be cleaned and inspected properly. Hence, the motor should be capable of running at inching speeds of 10-25 m/min as long as a particular button is pressed.

7. Each section should be able to run at the crawling speed of 10-25 m/min for running in felts, wire and heating up of dryer cylinders.

8. Smooth and quick starting of the sections, without causing excessive starting current, are required.
9. Control system employed should be flexible in nature.

10.2.4 Types of Drives Used for Paper Machine

There are two types of drives employed for making paper from pulp—line shaft drive and sectional drive.

Line shaft drives: In this form of drive, the various sections of the paper machine are driven from a line shaft running the full length of paper machine. Cone pulleys and belt-combinations drive the various sections of the machine from line shaft through right angled gear reductions.

Normally, electric motors are employed to drive the transmission shaft. Both ac and dc drives can be used for practically loss-less speed control.

In ac drive, only the ac commutator motor with shunt characteristic is of use for obtaining an economic speed control system. Its advantage is in the possibility of connecting the motor directly to a three phase supply, thereby eliminating an ac-dc converter, which would be required if a dc drive were used. However, the speed of an ac commutator motor depends on load and, hence, its use as a paper machine drive with stringent requirements of constant speed is no longer advisable. Also, its speed range (usually of the order 1 : 3) as well as the power required greatly affects the size of the motor. The open loop speed control of the ac commutator motor is sluggish in comparison with a dc drive, as speed is varied by adjusting an induction regulator and shifting the brush rocker.

In dc drives, the speed of the paper machine is controlled by varying the armature voltage of a separately excited dc motor. The variable dc voltage is obtained from alternating voltage supply by means of either rotary converters or static converters.

Sectional drives: In sectional drive each section of the paper machine has its own electrical motor. All the motors are operated from a common supply bus. By varying bus bar voltage, the speed of the paper machine can be controlled. By adjusting the field excitation of any motor, it is possible to vary the speed of that particular motor with respect to other motors.

10.2.5 Comparison Between Line Shaft Drive and Sectional Drive

The requirements of a line shaft drive are not as stringent as those of the sectional drive. It has a number of major limitations as compared to the sectional drive system. The lower cost of the electrical equipment of a line shaft drive system is offset significantly by the additional cost of the mechanical equipment required and its maintenance.

Draw: In a line shaft drive, the draw (speed difference between sections) is generally achieved as follows: The individual sections are coupled to the line shaft by means of cone pulleys and belts. By shifting the belt position, the section speed can be trimmed with respect to the master speed. The greatest problem in setting the draw is the slipping of belts. If the belt starts slipping the draw of the particular section gets affected and there is no automatic corrective action.

In sectional drives, the draw is obtained by trimming the speed reference to the sections and any variation in draw due to any reason is automatically corrected by means of closed loop feed back control systems.

Repeatability of draw settings: In a sectional drive the draw settings are done by calibrated accurate potentiometers. Therefore, the draw settings which are quite critical can be accurately recorded and reproduced, when the particular grade of paper is produced again.

In line shaft drive it is difficult to achieve the above.

Loop removal: In sectional drives, slight increase in the speed of a section is easily obtained, without disturbing the original draw settings for removal of loop.

In line shaft drives the loop removal has to be done by changing the draw settings. This invariably results in over correction. For example, assume that the paper has started forming a loop between dryer and calender. If the operator speeds up the calender to get the correct value, whatever loop that has already formed will continue to remain without reducing or increasing. The operator, now speeds up more and more until the loop is fully taken up and finds that the paper has become tight after some time. Then, he has to reduce the speed. This process goes on until the correct draw setting is obtained.

Rate of loop removal: In sectional drives, the rate of loop removal can be made constant at all operating speeds by suitable circuitry.

In the line shaft system, for a constant rate of loop removal, the belt shifting will be different at different operating speeds. At lower speeds the shifting will be more compared to that at higher speeds.

Independent operation of sections: With sectional drives, any section can be started and run at any required speed. For instance, during starting the wire section will be switched on for 'crawl' in order to do washing, whereas the dryers will be running at a higher speed for initial warming up.

In line shaft system, all the sections will necessarily have to run at the line shaft speed.

Crawl: While paper is being made, one of the sections may develop a minor trouble and it may have to be stopped. After the fault has been rectified, it may have to be put on 'crawl' for check up. With sectional drives this is possible but not so with line shaft.

Inch reverse: During threading, paper can get jammed in the dryer sections. To remove the paper it is essential to have 'inch reverse' facility. Line shaft system cannot offer this feature.

Electrical interlocks for safety: With sectional drives, each section can have its own set of devices that permit starting and safety interlocks. The former do not allow the sections to be started until all the necessary safety conditions for that particular section are fulfilled. Also, if any fault develops during running, the safety devices will trip only that particular section.

With line shaft system, human negligence can lead to major accidents. An operator can engage the clutch of the drive under the most dangerous conditions.

In the case of sectional drives, overloading in a particular section can be sensed and the drive can be shut off, if overloading is beyond the safe limits. Whereas, with the line shaft system overloading in a particular section has negligible effect on the drive motor and, hence, the overloaded section will be continued to be driven until the machine breaks down.

10.3 Cement Mills

10.3.1 Cement Making

The raw materials for producing cement contain lime and silica as main components and alumina and ferric oxide as fluxing components. The limestone mined from the quarries is crushed and transported to the plant by dumpers wagons, trucks or ropeways depending on the area and distance involved. In fact, if the quarry is within 1-2 km from the plant, the crushers might be located right next to the plant and in the line of supply of the limestone. The crushed limestone together with the required proportion of corrective additives like clay bauxite, iron ore etc. is ground in grinding mills. The fine dry powder coming out is homogenised in silos by passage of air from bottom and through the medium. It is then fed into the kiln, which is the heart of the cement plant, for producing cement clinker at high temperatures. If the kiln receives finely ground and precisely composed dry feed as mentioned above, the cement plant is called as a dry process one. In wet process, the raw materials are ground with water to produce a slurry before entering the kiln feed tank. Dry process is preferred to wet process because less fuel is required by such kilns. Wet process is necessitated sometimes, since, certain materials contain so much water that adding a little more water and using wet process is better than trying to dry the raw materials. The clinker coming out of the kilns is air cooled in special types of coolers and then transported to the storage. After aging in storage for at least three days the clinker, mixed with the right amount of gypsum is fed to the cement grinding mills and ground to required fineness. The cement is stored in silos, drawn for packing in gunny bags and despatched by wagons or trucks to the dealers.

10.3.2 Types of Drives

The driving motors used in the cement industry can be broadly classified as follows:

(i) Raw mill and cement mill drives
(ii) Kiln drives
(iii) Crusher drives
(iv) Waste gas fan drives
(v) Compressor drives etc.

Raw mills and cement mill drives: Slipring induction motors of 6.6 kV are widely used. In order to improve the power factor of the line current drawn, high voltage capacitors of adequate reliability and automatic capacitor control switchgear and circuit breakers are to be used. Even after adding the price of the capacitors and the control gear, the slipring motor is cheaper than the synchronous motor of the same rating. Liquid resistance starters are, usually employed to start the motor and to bring it upto full speed. Gear boxes are also attached in order to get the desired mill speed of about 15 rpm.

From the point of view of voltage dips during starting, the starting current of mill drives for large cement plants is normally restricted to 1.75 times the full load current.

The starting torque for the mill motors for large cement plants is limited to 125 per cent of the rated torque and the pull out torque is restricted to nearly 240 per cent of the rated torque.

Normally, the motors should be able to withstand 50 per cent overload for one minute occurring four times per hour at equal intervals. The motors for such drives are generally designed for a duty cycle of three consecutive starts from cold condition and two consecutive starts from hot conditions per hour against full load.

Twin drives: Due to the large ratings (above 3000 kW) required for the raw and cement mill drives and due to the limitations in the availability of large size gear boxes and motors, twin drives are employed in these mills. The two motors have to be more or less identical to each other and so also their liquid resistance starters.

Gearless drives: In developed countries, gearless drives are being increasingly used for large mills. The rotor is shrunk on to the mill and the air gap between the rotor and the overlapping stator is maintained by levitation using a sophisticated electronic closed loop control. The supply frequency is rectified into dc, which is then inverted to ac of a much lower frequency so as to provide a mill speed of approximately 15 rpm. This arrangement completely dispenses with the gearbox, which is normally the source of maintenance problems. These type of drives would become economically viable in a few years, when power diodes and thyristors would be available in plenty at a much lower cost than those prevalent today.

Kiln drives: The rotary kiln is an indispensable part of a cement plant. There are different types of rotary kilns depending on whether the cement is manufactured by means of wet or dry process. But, in general, they are tubular, slightly tilted from the horizontal and have a ring gear fitted around them, which engages with one or two pinions. Each pinion drive shaft is driven by a variable speed motor.

In early years, variable speed ac commutator motors were employed for kiln drives. However, due to requirements of higher output ratings and of speed range in excess of 1:2, the commutator motor became an expensive proposition in addition to maintenance problems. This was followed by

Ward-Leonard drives. On account of higher capital cost, lower efficiencies and greater maintenance problems as compared to thyristor controlled dc drives, the Ward-Leonard drive has been superseded by the dc motor with static supply.

The rating of the motors used for driving the kilns vary from 100-1000 kW. The maximum speed of the kiln is about 1 rpm and the kiln motor has to be designed for a speed range of the order of 1:10. The starting torque required may be between 200 per cent to 250 per cent of full load torque. The motors are also specially designed to pick up speed at full load within the normal time of 15 seconds. Quite often, kiln motors have to cater to overloads to the tune of 200 per cent to 250 per cent for small periods of time. The motor and control equipment have also to be specially designed for inching and spotting of the kiln during maintenance and routine checks.

Twin motor dc drives: To cope up with increasing kiln capacities, the modern trend is to use twin motor dc drives for kiln application. In this case, two dc motors with separate pinions drive the same gear wheel at the periphery of the kiln drum. Although this arrangement has certain advantages for the designer of electric motors, it does involve extra expenditure on the electrical side.

The twin motor drive system must be designed such that motive power is supplied in equal parts by the two motors to prevent overloading of either one of them or its mechanical transmission system. The speed of the two motors must necessarily be the same because they are coupled through the gear system. This can be achieved either with series connection of the two dc motors or with parallel connection by means of a closed loop control system.

Crusher drives: The motors used in crushers are of the slipring type. Stalling considerations play a very important part in the design of these motors. Normally, the motors are designed to withstand locked rotor current during running, without any external resistance introduced in the rotor circuit, for one minute. This is quite important, since very often the crushers tend to get jammed, when a big sized boulder gets trapped between the jaws of the crushers. Generally, the starting torque for such drives is limited to 160 per cent of the full load torque and the pull out torque is limited to 200 per cent to 250 per cent of the full load torque. The motors are also normally designed for 15 per cent overload for 15 seconds and 20 per cent overload for 10 seconds taking into consideration the adverse loading conditions encountered in practice.

Fan drives: The motors used are of the slipring type with a speed variation, generally, between 1000 and 750 rpm. The cast iron grid resistance controllers are normally used for starting and controlling the speed of these drives. As the motors are located outdoor or in semi-outdoor locations totally enclosed motors of TEFC are employed. The starting torque is around 120 per cent of the full load torque and the pull out torque is between 200 per cent and 250 per cent of the rated torque. The sliprings and brush gears are totally enclosed and kept external to the motor enclosures for ease in maintenance. Due to higher frequency and finer control, thyristorized dc drives are replacing slipring motors in fan drives too.

Compressor drives: The air compressor motors are normally of smaller ratings varying between 300 kW to 450 kW. Generally, the compressors are started on no load. Depending on specific system studies, either a squirrel cage or a slipring motor can be used. The enclosures are usually of TEFC type and the speeds vary between 1000 and 750 rpm.

10.4 Textile mills

From the stage cotton is picked in the field to the stage it leaves the mills as finished cloth, it undergoes different processes, viz., cotton to slivers, spinning, weaving and finishing.

Cotton to slivers: The process of separating the seeds from cotton is known as ginning. Ginning mills are usually located in the cotton growing areas. Bales of ginned cotton are first transported to the textile mills. There, they are opened and the impurities are picked up and removed in the blow room. After further opening and cleaning, cotton is transformed into laps and fed to cording section. Here it is opened completely and is converted into slivers. The slivers are gathered in cans and then processed on a drawing machine, which makes then uniform by straightening the fibre. The slivers are then changed into lap form before feeding them for combing, which parallels the fibre and upgrades it.

Spinning: The sliver at this stage is in a fragile condition and is also bulky. After reducing the diameter in two or three stages, it is processed on 'speed frame', which makes it suitable for final spinning. Due to twisting a continuous yarn of sufficient strength is produced during spinning. This yarn is wound on bobbins located in cone winding machines.

Weaving: Before the yarn is actually woven, it is 'warped', i.e., made into a uniform layer. Weaving consists of joining two sets of threads, one which extends throughout the length of the fabric and the other whose threads go across. This process is done in a loom.

Finishing: This consists of a number of processes like bleaching, dyeing, printing, calendering, stamping and packing. The impurities like oil and grease are removed and the fabric is made white during bleaching. Dyeing involves giving a colour or shade to the cloth. Printing produces designs and patterns in multicolours.

10.4.1. Motors Used for Different Textile Processes

All machines used in accomplishing the different processes described above require electric motors as their drives. Special environmental, operating and drive conditions demand specially designed motors for textile industry.

Loom motors: In order to accomplish the 'pick up' process in a short time, the starting torque of the loom motor should be high. The loom being essentially a reciprocating mechanism causes both torque and current pulsations. Also, loom motors are subjected to frequent starts and stops. These result in a higher temperature rise and is taken care by having good thermal dissipation capacity of the motor.

Loom motors are either totally enclosed or totally enclosed fan cooled, three phase high torque squirrel cage induction motors. Presence of lot of fluff in the atmosphere requires a smooth surface finish of the housing and end shields so that the fluff does not get collected on the surface of the motor. The insulation of the motor must be able to withstand high moisture content.

The ratings of the motors used for driving looms for light fabrics such as cotton, silk, rayon, nylon etc. are 0.37, 0.55, 0.75, 1.1 and 1.5 kW, while those of the motors used for making heavy fabrics (wool and canvas) are 2.2 and 3.7 kW. They are usually of 6 or 8 poles.

Card motors: The general requirement of card motors is almost similar to that of loom motors except that the former are required to have a very high starting torque and must be able to withstand a prolonged starting period. Both the above requirements for the card motor are due to the very high inertia of the card drum. Once the drum is started, the operation is continuous and uninterrupting, unlike that of a loom, where frequent starts and stops are involved.

The commonly used drives for card motors are again totally enclosed and totally enclosed fan cooled three phase high torque squirrel cage induction motors. The usual ratings of motors for cards of light fabrics are 1.1 and 1.5 kW and those for cards of heavy fabrics are 2.2, 3, 3.7 and 5.5 kW. Here again, the preferred synchronous speeds are 750 and 1000 rpm.

Spinning motors: For good quality spinning, it is essential that the starting torque of spinning motors should be moderate and the acceleration should be smooth. If the starting torque were low, the tension of the yarn would be insufficient and hence the yarn would get entangled and break. If the starting torque were high, the acceleration would be high and the yarn would snap.

In general, three types of drives are used for spinning frame operation: single speed motor, two-speed motor and two motor drive.

Normally, a 4 pole or 6 pole squirrel cage induction motor is used as single speed drive.

In order to maximize production with minimum breakage, two speed motors (4/6 or 6/8 poles) are used. Although these motors would be larger in size and costlier, the increased production may compensate for the additional initial outlay.

In case of two motor drive, two separate motors are used to drive the common pulley of the ring frame. Although this drive is costlier and requires more space, it has the following advantages:

(i) Any desired speed differential can be got by adjusting the speed ratios.
(ii) The tension of the yarn can be adjusted independently.
(iii) Production can be continued even when one of the motors fails.

Totally enclosed fan cooled motors ranging from 5-30 kW are used as spinning motors.

From what has been discussed above, it is clear that some form of controlled-torque starting of textile machinery drives is imperative. Proper starting will minimize yarn breakage, improve the quality of the product and increase the life of the machine. Less down time obtained will also increase production.

The most common electrical method of controlled-torque starting involve the use of standard squirrel cage motors and different methods of applying reduced voltage to the motors during a selected starting period. One of the most effective methods of controlling the stator voltage of motors used in textile industry is that of using series reactors. This method gives stepless, closed transition increase to approximately full speed. Since the motor terminal voltage is a function of the current drawn from the line, during acceleration the motor voltage will increase as the line current decreases, resulting in greater accelerating energy at the higher speeds and no significant pull on the yarn being processed. Reactors with fixed tappings may be used to get variety of starting torques. Variable iron-core inductors offer almost infinite choices of starting torques. Since the normal starting time is limited to a few seconds, the effect of poor power factor during starting on the overall plant power factor is not appreciable. Although primary resistor starting has all the advantages of reactor starting, it has not proved practicable for textile plants because of the possible fire hazards, when the resistors get overheated during frequent starting.

With the ever increasing use of solid state devices, nowadays, ac regulators are being used to provide control of starting torque. Fig. 9.4 shows the commonly used circuit. Through a tachometer speed feedback signal, machine starting can be automatically controlled by presetting the reference

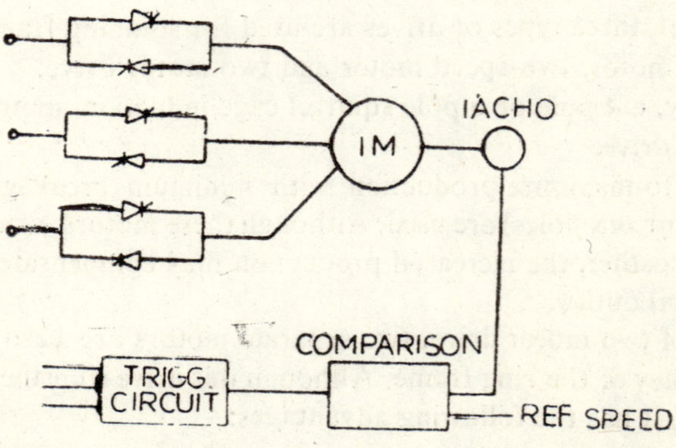

Fig. 10.4. Solid state stator voltage control of textile motor.

input signal from a potentiometer. Thus, a large variety of torque-time machine starts is possible from the same controller without any tap changing or adjustment of air gap of reactors.

10.5 Sugar Mills

In sugar mills, a centrifuge is used to separate out sugar crystals from the syrup by the action of centrifugal forces. The duty cycle of a centrifuge motor is as shown in Fig. 10.5. It consists of starting the motor to get a speed of about 200 rpm, charging the syrup with the motor disconnected from the supply, an intermediate spinning of the centrifuge at about 500 rpm, spinning at a speed of about 1000 rpm and braking of the motor successively at different speeds till a ploughing speed of about 50 rpm is obtained. As seen earlier, in chapters five and six, lot of energy will be wasted in both accelerating and braking the centrifuge motor. In order to reduce the energy lost during acceleration and retardation of the motor, the centrifugal action is performed at different speeds.

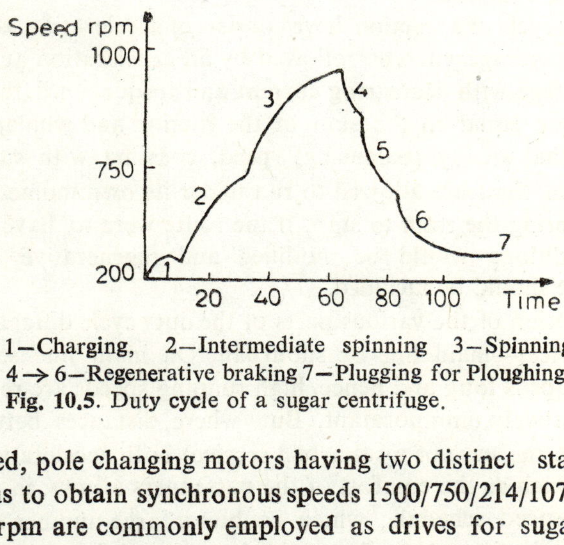

1—Charging, 2—Intermediate spinning 3—Spinning
4 → 6—Regenerative braking 7—Plugging for Ploughing
Fig. 10.5. Duty cycle of a sugar centrifuge.

Four speed, pole changing motors having two distinct stator windings that enable us to obtain synchronous speeds 1500/750/214/107 rpm or 1000/500/214/107 rpm are commonly employed as drives for sugar centrifuge. They not only are capable of providing the desired fixed speeds of operation, but also of returning a portion of the energy back to the supply during braking by means of regenerative braking accomplished by switching over to a higher pole operation from a lower pole one. In order to satisfy the duty cycle shown in Fig. 10.5, first the 28 pole winding is switched on to get a speed of about 200 rpm. The supply is cut off as the charging starts and as soon as it is over, the intermediate spin speed of about 450 rpm is obtained by energizing the 12 pole winding. The final spin speed near to 950 rpm is attained by switching over to the 6 pole winding. Once the final spin speed of the centrifuge has been reached, supersynchronous regenerative braking is done by connecting the supply to 12, 28 and 56 pole windings successively to bring the speed down to near about 500, 214 and 107 rpm respectively. Ploughing speed of about 50 rpm is accomplished by plugging the 56 pole motor. Automatic control of the entire duty cycle is achieved by feeding the control equipment from the output of the tachogenerator, which is directly coupled to the motor driving the centrifuge.

Motors used for driving centrifuges have their own special features in construction. They are vertically mounted so as to be coupled with the shaft of the centrifuge. The motors may have larger air gap to take care of the possible rotor oscillations about the vertical axis. Since the motors operate in a humid environment, the insulation used must be humid proof. Due to frequent switching of the motor at different windings wound for different poles, wide fluctuations in motor current occur during the duty cycle of a centrifuge motor. To protect the motor from overheating, thermal elements, which operate a few degrees below the maximum permissible temperature of the winding, are embedded in the windings. These elements, often, give visual and audible warning signals so that the particular duty cycle may be completed.

10.6 Electric Traction

The duty cycle of a traction drive consists of acceleration at the maximum permissible average current followed by an acceleration at maximum permissible voltage with decreasing current and torque until the motor develops a torque equal to the sum of the friction and windage torques, free running at that steady (balancing) speed, coasting with supply to motor 'cut off' and the train allowed to run under its own momentum and lastly braking to bring the train to stop. If the route were to have gradients, the above conditions would be modified and regenerative braking down the gradient should be included.

The duration of the various parts of the duty cycle differ according to the type of services—main line or suburban. On main line service, distance between stops is long and hence, high running speeds are required. Acceleration is relatively unimporatant. But, where distances between stops are small (as in the case of a suburban service) both acceleration and braking must be as high as the comfort of the passengers allows, in order to achieve a fast speed schedule, which enables a large number of trains to run within a certain time. Also, in a suburban type of service, there will be practically no time for free running and acceleration is followed immediately by coasting and braking. The speed time curves corresponding to both types of services are shown in Fig. 10.6. The area enclosed by the speed time curve gives the distance travelled by the train and the slope at any point on the curve towards abscissa indicates the acceleration or retardation at that instant of time.

In order to provide necessary linear and angular acceleration to the train mass and to overcome the gravity component of the weight of the train, the wind and frictional resistance of the train and any curve resistance present, a traction unit has to develop a certain force called tractive effort at the

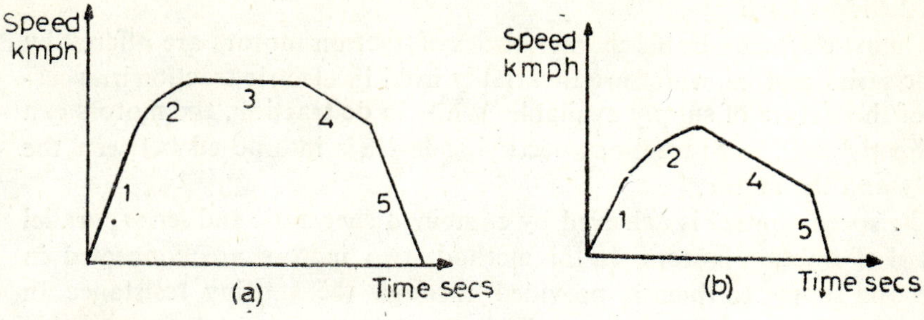

1. Acceleration at constant current
2. Acceleration at constant voltage
3. Free running
4. Coasting
5. Braking

Fig. 10.6. (a) Speed-time curve for mainline service.
(b) Suburban service.

wheel rims. This force should be maximum during the accelerating period. During the free running period, although the train will be moving at a high speed, the tractive effort required is not much since it has to overcome only gravity component and track resistance. The motors that provide the desired tractive effort are known as traction motors.

10.6.1 Requirements of Traction Motors

Since the tractive effort at the rim of the driving wheels is directly proportional to the output torque of the motor, the traction motor must develop high starting torque and reduced torque at high speeds.

In order to provide large tractive effort, more than one motor is usually required and they are connected in parallel. The speed torque and speed current of these motors should be of such a nature that for minor differences in the speeds of rotation of different motors connected in parallel, the differences in the torques developed and currents drawn are quite small. Also, if the speed torques characteristic is of such a shape that the torque developed is inversely proportional to the speed, thereby making the power demanded from the supply more or less constant, the motors will have an inherent protection from overloading.

Motor should be amenable to simple methods of speed control and braking, particularly regenerative braking. Traction motors should be designed to have a large power/weight ratio to limit the weight and size of motors; they should be robust in construction and be of high efficiency.

Due to the possibility of a heavy inrush current at starting, fluctuations in supply voltage is almost a normal feature in traction practice. Also, temporary interruption of power supply is also likely to be present. Therefore, traction motors must be capable of withstanding the voltage fluctuations and temporary interruptions in supply.

10.6.2 Drives Used

Almost all the desirable characteristics of traction motors are offered by the dc series motors, which are invariably used in electric traction irrespective of the nature of supply available. While in dc traction, the motors can be directly used, in ac traction a rectifying device is interposed between the supply and the motor.

The speed control is achieved by combined rheostatic and series parallel control of traction motors. In this method, two motors are connected in series and supply to them is provided through the starting resistance in series with the motors, in order to limit the starting current to a reasonable value. Then, the added external resistance is gradually cut out until two motors only are working in series. At this position, each motor is receiving one half of the supply voltage and therefore runs at one-half of the rated speed. To achieve full speed, the motors have to be disconnected and then reconnected in parallel, again fed through a starting resistance. This resistance is slowly cut out and the motors alone are left to operate in parallel. The change over from series connection of motors to parallel connection should be accomplished in a number of steps in such a manner that there would be minimum disturbance to the supply system.

Very often, four or six motors are employed to provide the required tractive effort. With four motors present, three types of connections are used—at first all the four motors are connected in series with one-fourth of the supply voltage fed to each motor; two sets of motors in parallel, each set consisting of two motors in series and each motor being fed with one-half of the voltage and all the four motors in parallel with full supply voltage across each motor. Similar interconnections can be made with six motors also in order to have different speeds of operation.

The only serious disadvantage of using dc series motors in electric traction is their inability to operate stably during regenerative braking. For example, a small increase in line voltage will cause a reduction in the magnitude of the current supplied during regeneration. This reduced current passing through the series field winding will create less flux and thereby less generated emf. This process will continue till the emf generated and the supply voltage are equal and regenerative action stops. Similarly, a small decrease in line voltage will cause excessive currents during regeneration. Hence, before using series motor for regenerative braking, it is necessary to excite the field windings separately and use stabilizing circuits.

Advances made in the technology and fabrication of solid state devices have enabled the use of both thyristor converters and choppers for driving the dc motors used in electric traction with increased efficiency, reliability and response.

10.7 Coal Mining

Electric motors are used widely in all areas of mining and installation. They can be classified into two basic groups. Motors belonging to the first group are used as auxiliaries to the basic function of mining. They include drives for compressors, fans, pumps, conveyors and hoists. The second group consists of motors at the mine face that are directly employed in cutting and removal of coal. These motors are used on continuous miners, cutting machines, drills, loaders, shuttle cars and extensible face conveyors, where they drive cutting chains, ripper bars, boring arms, drills, pumps, loading arms, conveyors and traction drives.

Auxiliary motors and mine face motor differ considerably from each other. The former are usually modifications of general purpose industrial motors, while the latter are specifically designed for specific needs. The duty of the auxiliary motors is often well defined and steady, while that of face motors consists of random loading and has a number of high-shock loads. Both electrical and mechanical demands made on face motors are very much greater than those made on auxiliary motors, and, hence, they are specially designed. Due to limited space available at the mine face, the mine motor is restricted by space, so that the external geometry is chosen to fit into the available space on the mining machine. Normally, auxiliary motors are fixed to the floor, while the face motors are mounted on mining machines.

10.7.1 Motors Used for Different Functions

Coal drills: The drills used for drilling in coal face are light in weight and rotary in type. They usually have a 1 kW, 125 volts, 2860 rpm, 50 Hz induction motor with a flameproof enclosure. The motor rotates a drill rod, with a tipped bit attached to its end at a suitable speed by using a gear.

The coal drill motor is short time rated. Sometimes, high frequency motors (150 or 200 Hz) are used with the help of a frequency changer housed in the drill panel. Such drills will have reduced size and higher power ratings.

Mine Winders: A colliery winder is used to raise coal, to lower and raise men and to lower or raise other loads. Fig. 10.7 shows the basic components of an electric winder. Wound rotor induction motor with water rheostats are, generally, used for driving the winder. DC rheostatic braking is employed initially and mechanical braking is used only towards the fag end of the winding period: For higher capacities dc motors supplied from a Ward-Leonard generator are used to drive mine winders, since precise control of drum speed and position is essential for accurate decking. Further, for large capacities, the maximum KVA demand on the supply system and the energy costs are less than that of equivalent capacity slipring induction motor drive, especially when used for deep shafts. The recent trends are to use thyristor converters to supply variable dc voltage.

Haulage (*Mine-winches*): The transportation of tubs along rails from the filling place to the pit bottom, where the tubs are run into the cages for conveyance by the mine winder to the surface is called haulage.

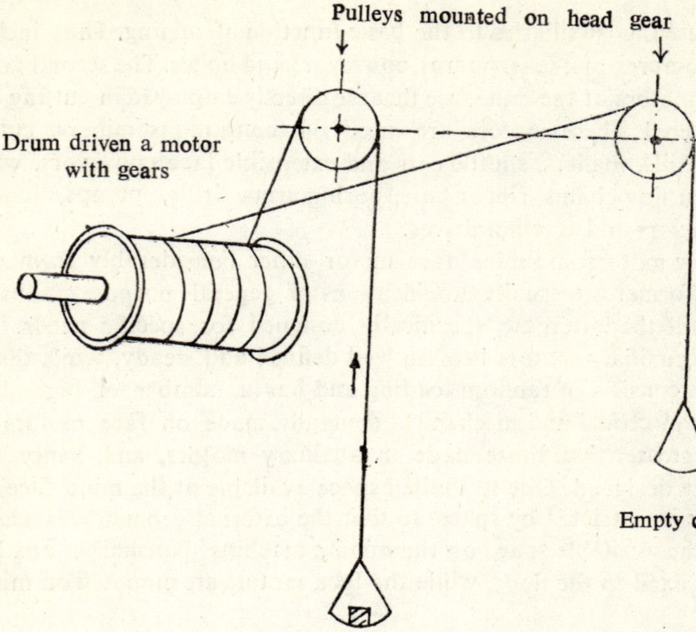

Fig. 10.7. Colliery winder.

Normally, a squirrel cage motor with high pull out torque is preferred because of its simple and robust construction. The drive is taken up gradually using a clutch so that the motor need not develop a high starting torque. However, often the loads require different speeds at different stages of haulage. Further, the motors are subjected to frequent starts and stops and also to braking and reversals. Hence, a wound rotor motor with drum resistance controller is used. The drum controller is set to introduce all external resistance on starting and manually adjusted through successive contact steps to accelerate the motor. It is also used to achieve different operating speeds. In order to ensure that the drum controller is returned to the initial position before starting, an interlocking arrangement is provided. The drum controller and the rotor resistances are separately mounted in flame proof enclosures.

Mine ventilation fans: There are two different types of fans used—axial flow and radial flow (centrifugal). In the former the air flow is axial, while the latter produces pressure through the centrifugal force of a rotating mass of air. They exhibit different characteristics and a thorough study of the proposed mine ventilation system has to be made before choosing the correct type of fan.

The speed of an axial fan, normally, is in the range of 600-1000 rpm and that of a centrifugal fan lower than this. Although squirrel cage motors having large number of poles can be used directly, high speed motors of 4 or 6 poles with a reduction gear are widely employed, since they have the advantage of greater efficiency and facility to operate the fan at different speeds by changing the gear ratio. The motor is usually connected to the 3.3 kV supply system of the colliery.

Pumps: Pumping water is, generally, carried out in two stages. Water from all sections is first brought to the main pump, which, usually, is located near the mine entry. From there, it is pumped to the surface.

Colliery pumps are of two types—triple ram type driven by a high speed motor with a gear and the centrifugal type directly coupled to the driving motor. The ram type pump needs a starting torque of about twice the full load value and, hence, slipring induction motors are used as drives. However, the centrifugal pump requires the motor to provide with only about 40 per cent of full load torque at starting, if the delivery and bypass are closed. Therefore, high torque squirrel cage motors, which do not have any maintenance problems, are adequate.

Working surroundings is the chief factor which distinguishes electrical motors used in mining from those used in general industry. Continuous extraction of coal necessitates frequent movement of electrical equipment from one site to another. The enclosures must be designed to withstand the ordeal of underground transportation and also to safeguard the motor from falling debris of coal, stones or water. They must also be flameproof so that the motors can be safely operated in the hazardous gassy and dusty atmosphere of the mine face areas without causing ignition of fires or explosions.

10.8 Machine Tool Applications

Squirrel cage induction motors are the most widely used ones for most machine tool drives due to its simplicity, reliability, low cost and minimum maintenance requirements. Most of the requirements of general purpose machine tools are easily met by these induction motors together with simple manual controls. Gear boxes and stepped pulleys are used to transmit the power to cutting tools or work piece at different finite number of speeds. The use of gear mechanism produces vibrations and noise, hence affects the accuracy of machining.

Better machining timings and surface finish are achieved, if stepless speed control is adopted. Electro-hydraulic and electromagnetic controls are employed to get the desired smooth speed control. With the advent of thyristor controlled drives, these methods have been superseded. At present, in our country, the use of thyristorized drives is limited to special machine tools only, as the cost of these drives is sufficiently high.

For use such as internal grinding spindles for horological applications, drilling of printed circuit boards, etc., high speed motors are generally employed. These are usually high frequency, three phase induction motors operating upto speeds of 1,80,000 rpm and giving fairly high output powers in small sizes. Typical ratings are 1.2 kW at 1,20,000 rpm for high speed grinder drives and 0.5kW at 1,50,000 rpm for high precision horological applications.

Certain machine tool applications require two or more fixed speeds. Poel changing motors with single or two stator windings and pole amplitude modulated motors are used to achieve the desired speeds.

10.8.1 Drives for Numerical Controlled Machines

With the advances made in the digital processing area, numerical controlled machines are slowly replacing conventional machine tools. With conventional machine tools the machine utility is quite low, whereas with the numerical controlled machines the utility could be five to six times more. Taking into account the operational inefficiencies, coordination required, high rate of rejections, number of stage inspections, etc., one numerical controlled machine possibly can do the job of perhaps four to six conventional machines. As regards cost, with the development of indigenous drives and controls, numerical controlled machines may become economical soon.

Essentially the motors for numerical controlled machines are required for spindle drive and number of axes drives. The number of motors in a specific numerical controlled machine could be from two to six depending on the type and sophistication of the machine. The principal requirements for these drives are very fast response, wide speed range, very low inertia, precise positioning accuracies, severe duty cycles, low vibrations, better thermal capacity, low maintenance cost, etc. Depending upon the control system, the drive motors for axes are required to be provided with a resolver or a digital encoder, coupled to the motor shaft for position feedback along with a tachogenerator integral with the motor, for velocity feedback. The motor ratings for the spindle drive will be usually upto 25 kW and 3000 rpm and for axes drive upto 10 kW and 5000 rpm.

Electronically commutated dc motors or inverter-fed induction motors are used for spindle drives, whereas stepper motors or high energy rare-earth permanent magnet dc servomotors are employed for axes drives.

10.9 Petrochemical Industry

10.9.1 Pump drives

The petroleum and chemical industries use fluid handling equipment extensively. Pumps have been driven by induction motors with control of flow, when desired, accomplished by throttling valves. It has been estimated that thirty percent or more of the electrical energy used by a typical refinery is wasted as a result of the throttling effect of control valves. Howver, adjustable-speed pumping utilizing multispeed induction motors or inverter-fed induction motors offer substantial economic benefits based on reduced energy and maintenance costs. Since conservation of energy has become a must for all industries and variable flow applications are commonly present in many industries, the energy saving capability of adjustable speed drives has been discussed in detail below:

Figure 10.8(a) shows a typical pump characteristic describing the head

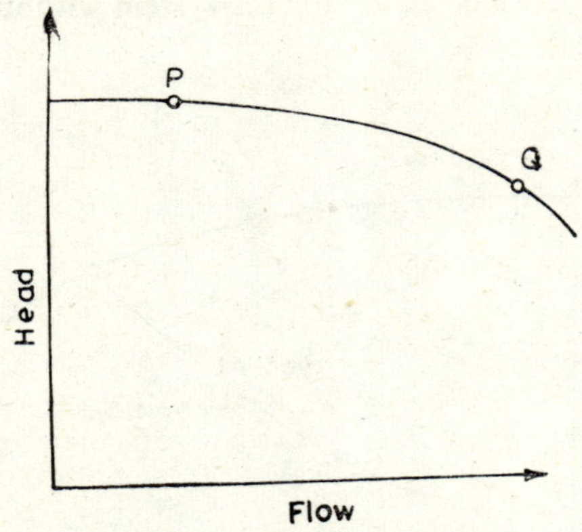

Fig. 10.8. (a) **Pump curve.**

(or pressure) versus flow variation of a centrifugal pump. This curve indicates that the pump will produce limited flow if applied to a piping system in which a large change in pressure is required across the pump to lift the liquid and overcome resistance to flow (as at point *P*). Higher flow rates can be obtained as the required pressure differential is reduced (as at point *Q*).

The system curve [Fig. 10.8(b)], which represents the characteristics of

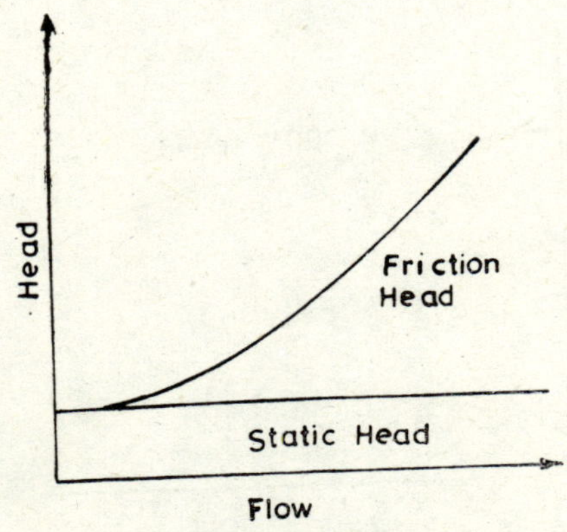

Fig. 10.8. (b) **System curve.**

the piping system to which the pump is applied, is required to determine where along the pump characteristic it will operate in a given application. The head required at zero flow is known as static head or lift. This shows how many metres of elevation the pump must lift the fluid irrespective of the rate of flow. The other component of head, called the friction head, increases with increasing flow. It is a measure of the resistance to flow (back-pressure) offered by the pipe and its accompanying valves, elbows and other system elements. The intersection of pump and system characteristics denotes the natural operating point for the system without flow control, as illustrated in Fig. 10.9.

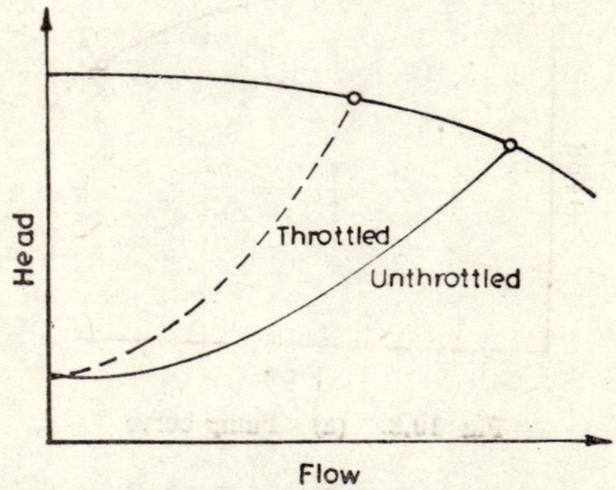

Fig. 10.9. Throttling valve control of flow.

If a control valve is introduced into the piping system at the pump outlet and partially closed, it reduces flow by increasing the friction in the system. The modified system curve and the new operating point are also shown in Fig. 10.9. It may be observed that the desired reduction in flow has been achieved but the head has been increased.

If the desired flow could be accomplished by modifying the pump curve rather than the system curve, as shown in Fig. 10.10, both flow and head

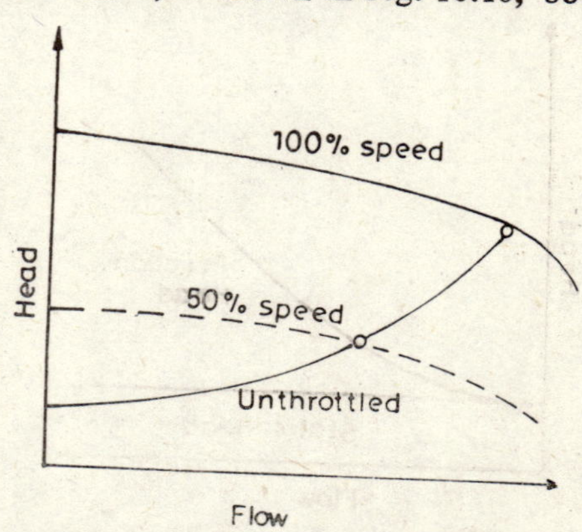

Fig. 10.10. Variable speed control of flow.

could be reduced together. Actually, the head required across the pump would be reduced rapidly relative to flow reduction. Reducing the speed of the pump results in change of the pump characteristic as shown. Since the pump output power is proportional to the product of head and flow for any fluid, it is obvious that there is a substantial reduction in required pump output power when the speed of the pump is reduced.

Variable-speed drives using solid state devices have much lower losses than others and, hence, substantial energy savings are achieved by using them for pumping applications.

10.9.2 Compressor drives

Important tasks within the production processes of the petrochemical industry include gas liquefaction, compression, refrigeration and heat recovery. Compressor drive systems constitute important components in plants for such processes. In the past, large high speed radial and axial compressors were driven mainly by turbines, but in recent years electric motors have been used increasingly. In addition to lower plant costs, the advantages include high operating efficiency, less pollution and more simple handling and maintenance. The majority of drives operate at a constant speed, using a four-or-six pole motor with step up gear. For large ratings, synchronous motors with cylindrical rotors having both excitation and starting windings (damper bars) are preferred. These motors have to be specially designed to take care of the oscillating torques produced during asynchronous acceleration due to the magnetic and electrical anisotropy in the rotor and special cooling conditions.

10.10 Miscellaneous Applications

Table 10.1 indicates the types of motors used for driving some more common applications.

Table 10.1 Guide for selection of motors for certain applications

Application	Special features required	Type of motor
Large ships	Large power demand	Synchronous or slipring
Small ships	Medium power demand	Slipring or dc
Centrifugal pumps	Normally started with closed valve. About 40-50 per cent rated torque required for starting	Standard squirrel cage
Reciprocating pumps	Normally started with closed valve; About 100-200 per cent rated torque required for starting	Slipring
Rotary compressors	Unloaded starting	Squirrel cage or slipring
Reciprocating compressors	Starting with partial load	Slipring
Centrifugal fans	Started with closed dampers or valves	Squirrel cage or slipring
Blowers	Starting with valves closed	Squirrel cage or slipring
Stirrer for liquids	Light starting duty	Squirrel cage or slipring
Cranes and hoists	Heavy starting torque, frequent starts and stops	High torque squirrel cage or slipring
Lifts	Smooth uniform acceleration, quiet running, frequent starts and stops	High torque squirrel cage or slipring
Line shafts, load conveyors, pulp grinders	Started on load; Normal starting torque of 200 per cent rated torque required	High torque squirrel cage or slipring
Flour and rubber mills, circular saws, chippers, beaters, crushers without flywheels etc.	High peak loads: A maximum torque of 200-250 per cent rated torque required; Frequent starts, stops and reversals	High torque squirrel cage or slipring
Crushers, punch presses, power hammers etc. all with flywheels	High peak loads; Heavy starting torque	High slip; High torque squirrel cage or slipring
Domestic appliances: mixers, sewing machines, vacuum cleaners	Simple speed control	Universal
Domestic refrigerator compressor, window airconditioners	Good starting torque	Single phase capacitor with direct on line start

Bibliography*

1. R.W Jones, *Electric Control Systems*, New York, Wiley, Second Printing, 1959.
2. G.J. Thaler and M.L. Wilcox, *Electric Machines: Dynamics and Steady State*, Wiley International Edition, New York, Wiley, 1966.
3. N.N. Hancock, *Electric Power Utilization*, London, Isaac Pitman, 1967.
4. A.E. Fitzgerald, C. Kingsley and A. Kusko, *Electric Machinery*, International Student Edition, New Delhi, McGraw-Hill Kogakusha, Third Edition, 1971.
5. M. Ge. Chilikin, *Obshiye Kurs Electroprivada*, Moscow, Energia, 1971.
6. L. Ba. Geyler, *Osnovi Electroprivada*, Minsk, Visheshaya Shkola, 1972.
7. G. De. *Electrical Drives and their Controls*, Bombay, Academic Book Ltd., 1970.
8. R.S. Ramshaw, *Power Electronics—Thyristor Controlled Power for Electric Motors*, London, Chapman and Hall, 1973.
9. J.M.D. Murphy, *Thyristor Control of A.C. Motors*, Oxford, Pergamon Press, 1973.
10. H. Partab, *Art and Science of Utilization of Electrical Energy*, Agra, Pritam Surat and Bros., Second Edition, 1975.
11. K.K. Mukherjee, 'D.C. drives and controls and applications in rolling mills', *Electrical India*, Vol. XIV No. 17, 15 Septemper, 1974, p. 13.
12. B.P. Sarkar, 'Electrical equipment for main drives of reverse cold rolling mills,' *Journal of the Institution of Engineers* (India), 56, Pt EL 2, October 1975, p. 57.
13. D.J. Thakur, 'Application of DC drives in the paper industry,' Tutorial on Rotating Machines, IEE, Bombay Branch, November 1980.
14. H.V. Wurgler, 'Electrical drives for rotary cement kilns,' *Brown Boweri Review*, **58**, 1971, p. 106.
15. J.R. Eliason and B.S. Fischer, 'Large adjustable speed fan drives including static converter development for cement plants,' *IEEE Trans. Ind. Appl.* Vol. 13, 1977, p. 557.
16. S.M. Gharse, 'Special features on rotating machines in cement plants', Tutorial on Rotating Machines, IEE Bombay Branch, November, 1980.
17. W. Harveydrake and H.L. Prescott, 'Controlled torque starting of textile machinery,' *IEEE Trans. on Ind. and Gen. Applic.*, **IGA-7**, 1971, p. 267.
18. M.S. Padbidri, 'Electric motors for textile industry,' *Electrical India*, **XVII**, No. 24, 31 December 1977, p. 35.
19. A.V. Wagle, 'Electrical drives for coal mining,' *Electrical India*, **XIX**, No. 16. 31 August 1979, p. 17.
20. T. Appaswamy and A. Bhargava, 'Some aspects on the application of electric motors in the machine tools industry,' Tutorial on Rotating machines, IEE Bombay Branch, November, 1980.

*The following list of references is by no means complete. It contains only a few books and papers which are extremely useful.

21. A. Kusko and S.M. Peeran, 'Brushless D.C. Motors using Unsymmetrical Field Magentization', IEEE Trans. Ind. Appl. Vol. IA-23, No.2, 1987, p. 319.

22. R.J. Blake, P.D. Webster and D.M. Sugden, 'The Application of GTOs to switched reluctance drives,' Second Internat. Conf. on Power Electronics and Var. Speed Drives, Nov. 1986, p. 24.

23. D.E. Bruskin, A.E. Zarokhovich and V.S. Khvostov, Elecktricheskie Mashini-2, Moscow, Vishshaya Shkola, 1979.

24. D.A. Jarc and J.D Robechek, 'Static induction motor drive capabilities for the Petroleum Industry,' IEEE Trans. Ind. Appl., Vol. IA-18, No. 1, 1982, p. 41.

Index